Contents

GCSE
Mathematics
Revision
Guide (HigherTier)

Brian Speed

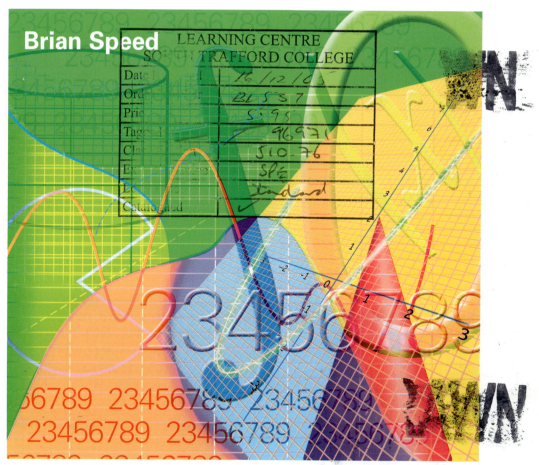

5.95

Philip Allan Updates
Market Place
Deddington
Oxfordshire
OX15 0SE

Tel: 01869 338652
Fax: 01869 337590
e-mail: sales@philipallan.co.uk
www.philipallan.co.uk

© Philip Allan Updates 2003

ISBN 0 86003 757 6

Printed by Raithby, Lawrence & Co. Ltd, Leicester

Introduction

About this guide

This book is written for students following a GCSE Mathematics Higher Tier course. It covers seven topics that are common to all specifications:

- Number
- Algebra
- Graphs
- Shape
- Pythagoras and trigonometry
- Geometry and construction
- Statistics and probability

As a revision guide, it does not replace your class notes but summarises the material that you need to learn. It uses worked examples to illustrate the processes involved in each topic, and provides you with a concise set of notes.

Each topic includes a 'test yourself' section. This is a series of short questions aimed at enhancing your revision. It will either confirm that you have understood the basic ideas in that topic or guide you to the areas where you need to spend more time.

A set of practice questions concludes each section. These are GCSE Higher Tier style questions written to help you prepare for what you might face in the exam.

Answers to all the 'test yourself' and practice questions are provided on pages 108–133. Answers to most of the practice questions are complete with working, to show you what you should have done to find the answer. Use this section wisely, that is, only after you have attempted the questions. Read the hints to see where you may have gone wrong and remember that it is OK to make mistakes — as long as you learn from them.

Preparation for the exam

- Start in good time. There is a lot to revise and you will need to go over some topics more than once.
- Make sure you know which topics are on your specification and be aware of your strongest and weakest areas. Spend that bit longer on your weaker topics but do not neglect your strengths.

- Do not simply read this book! Actively work through it. Attempt all the questions and test yourself thoroughly. The best way to revise mathematics is to do some; then do some more!
- Try the practice questions under exam conditions if you can. Then use the answers at the back to check how you got on.
- Don't be surprised if some topics or questions seem difficult. This is, after all, the Higher Tier. Tell yourself, 'I am quite capable of getting a good grade, otherwise I would not have been entered at this level.'

Command words used in the exam

There are some words used in questions that tell you how you should answer the question. These are outlined below — become familiar with them.

- **Write down** — you do not need to do any calculation. The answer should be straightforward.
- **Calculate** — you do need to perform calculations to reach the answer.
- **Evaluate** — this is another way of saying 'calculate'.
- **Find** — you *might* have to do some calculations to find the answer.
- **What is** — this is another way of saying 'find'.
- **Explain** — you need to show clearly why you are saying something.
- **Construct** — you must use a compass to construct the diagram.
- **Approximately** — you must give a rounded answer. A sensible rounding will be something like an integer or the nearest hundred. Use the context of the question to help you decide on this.

Hints for the exam

- You have met every topic on the exam paper, so you can attempt every question.
- Read each question at least twice. Most marks are lost by students not answering the actual question asked.
- Show your working if there is more than 1 mark for the question. If you get the answer wrong, there are still marks to be gained for your working.
- Look at the marks per question. This is a hint as to how much work needs to be done for the answer.
- If some major topics have not appeared in Paper 1, then they will almost certainly appear in Paper 2.

Number

There are certain skills that you will need to use only occasionally but there are others, like rounding off and percentages, that you will use often and on every exam paper. You need to be familiar with these topics or you could lose quite a few marks.

Basic percentage

To find A% of an amount P, calculate $(A \times P) \div 100$.

Example

8% of 40 kg is

$(8 \times 40) \div 100 = 3.2$ kg

Any percentage can be represented as a decimal by dividing by 100. For example:

$35\% = 35 \div 100 = 0.35$

Some common fractions expressed as percentages are:

$\frac{1}{2} = 50\%$ $\frac{1}{4} = 25\%$ $\frac{3}{4} = 75\%$ $\frac{1}{10} = 10\%$ $\frac{1}{5} = 20\%$ $\frac{1}{3} = 33\frac{1}{3}\%$

Percentage change

Percentage increase

To calculate percentage increase, change the percentage to a decimal (divide by 100), add 1, then multiply by the figure that needs increasing. For example, to increase £7 by 5%:

- change 5% to a decimal (0.05)
- then add 1 (1 + 0.05 = 1.05)
- then multiply by £7
  i.e. $1.05 \times £7 = £7.35$

Percentage decrease

To calculate percentage decrease, change the percentage to a decimal, subtract it from 1, then multiply by the figure that needs decreasing. For example, to decrease £8 by 14%:

- change 14% to a decimal (0.14)
- then subtract from 1 (1 − 0.14 = 0.86)

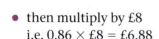

- then multiply by £8
  i.e. $0.86 \times £8 = £6.88$

Expressing one quantity as a percentage of another

To express one quantity as a percentage of another, set up a fraction and convert that fraction to a percentage by simply multiplying by 100. For example, to express £6 as a percentage of £25:

- set up the fraction $\frac{6}{25}$
- multiply by 100
- this becomes $(6 \div 25) \times 100 = 24\%$

Reverse percentage

There are times when we know a certain percentage and we wish to get back to the original amount.

Example

The 36 pupils who were absent represented 3% of the pupils in the school. How many pupils should have been at school?

- Since 3% represents 36 pupils, then 1% will represent $36 \div 3 = 12$ pupils
- So 100% will be represented by $(12 \times 100) = 1200$ pupils

Compound interest
added

In a bank account that pays compound interest, the interest is added to the original amount before the next percentage change is applied.

Example

James puts £80 in a bank that adds 6% compound interest at the end of each complete year. How much will he have in the account at the end of 3 years?

After:

1 year	$80 \times 1.06 = 84.8$
2 years	$84.8 \times 1.06 = 89.888$
3 years	$89.888 \times 1.06 = £95.28$ (to the nearest penny)

or quicker still:

$80 \times (1.06)^3 = £95.28$

Notice that the power is simply the number of years that compound interest is added.

Ratios

To divide an amount in a given ratio, you simply multiply the amount by the fraction of the ratio. For example, to divide £70 between Joe and Ally in the ratio of 2:3 we see that:

- Joe receives $\frac{2}{5}$

- Ally receives $\frac{3}{5}$

- so Joe receives $£70 \times \frac{2}{5} = £28$ and Ally receives $£70 \times \frac{3}{5} = £42$

If you know part of the information

Example

Two business partners, Sue and Trish, divide their profits in the ratio 3:5. If Sue receives £1800, how much does Trish receive?

- Sue receives £1800 which is $\frac{3}{8}$ of the whole profit

- so $\frac{1}{8} = £1800 \div 3 = £600$

- so Trish's share, which is $\frac{5}{8}$, will be $£600 \times 5 = £3000$

Fractions

Adding and subtracting

First change both denominators (bottom numbers) to be the same, which also changes the top numbers (the numerators). Then add or subtract.

Example

$$\frac{3}{5} + \frac{7}{8}$$

- Change the denominators to 40 (5×8), which means the fractions become:

$$\frac{3 \times 8}{40} + \frac{7 \times 5}{40}$$

$$= \frac{24}{40} + \frac{35}{40} = \frac{59}{40}$$

- This is top-heavy, so change to a mixed number:

$$1\frac{19}{40}$$

$$\frac{2}{3} - \frac{5}{8}$$

- Change the denominators to 24 (3×8), which means the fractions become:

$$\frac{2 \times 8}{24} - \frac{5 \times 3}{24}$$

$$= \frac{16}{24} - \frac{15}{24}$$

$$= \frac{1}{24}$$

To multiply

First cancel, then multiply across. For example:

$$\frac{5}{9} \times \frac{6}{25}$$

Cancel the 5 and 25 by dividing each by 5.

$$= \frac{1\cancel{5}}{\cancel{9}_3} \times \frac{^2\cancel{6}}{\cancel{25}_5}$$

Cancel the 9 and 6 by dividing each by 3.

$$= \frac{1}{3} \times \frac{2}{5} = \frac{2}{15}$$

To divide

Turn the second fraction upside down and then multiply. For example:

$$\frac{5}{6} \div \frac{2}{3}$$

$$= \frac{5}{\cancel{6}_2} \times \frac{1\cancel{3}}{2}$$

Cancel the 3 and 6 by dividing both by 3.

$$= \frac{5}{2} \times \frac{1}{2}$$

$$= \frac{5}{4}$$

which is top-heavy and simplifies to:

$$1\frac{1}{4}$$

Approximation

Find an approximation of 26.3×7.48.

To approximate the answer to this and other similar situations, simply round each number off to 1 significant figure (s.f.) and then do the calculation with the rounded values.

So, for 26.3×7.48 an approximation is:

$$26.3 \times 7.48 \approx 30 \times 7 = 210$$

Sometimes, especially when dividing, we round a number off to something useful at 2 s.f. instead of 1 s.f. For example:

$$49.2 \div 6.37$$

Since 6.37 rounds off to 6, let us round 49.2 off to 48, since 6 divides exactly into 48. So:

$$49.2 \div 6.37 \approx 48 \div 6 = 8$$

It is a good idea to estimate the answer to any calculation. You can then check your calculated answer against your estimate, to make sure it is sensible.

Limits of accuracy

Whenever we round off or approximate we immediately bring a slight error into the figures. For example, if I said my height is 168 cm, then I could be as small as 167.5 cm or as tall as 168.49999999 cm because:

(i) the smallest value that can round up to 168 is 167.5

(ii) the largest value that can round down to 168 is 168.49999999

These lowest and highest values are called the limits of accuracy.

All measurements are rounded off to some degree of accuracy. This determines the possible true values before the rounding took place. For example, a length of 23.6 cm is rounded to 1 decimal place:

- the smallest possible value is 23.55
- the largest possible value is 23.6499999999
  i.e. $23.55 \leq$ length < 23.65

Example

The sides of a rectangle are given as 12 cm and 9.5 cm. Calculate:

(a) the greatest possible area the rectangle could actually be

(b) the greatest possible percentage error in the calculated area

Answer

(a) The upper limit of both sides will give the largest area. The upper limits are just under 12.5 and 9.55 respectively. This will give an area of just under $12.5 \times 9.55 = 119.375$. Hence the greatest area will be just under 119.375 cm².

(b) The greatest percentage error is at the lower limit of each side, giving an area of $11.5 \times 9.45 = 108.675\,cm^2$.

The given lengths produce an area of $12 \times 9.5 = 114\,cm^2$

The error would be $114 - 108.675 = 5.325\,cm^2$

The greatest percentage error $= \dfrac{5.325}{108.675} \times 100 = 4.90\%$

Types of number

Rational and irrational numbers

The definition of a **rational** number is a number that can be expressed in the form $\frac{a}{b}$ where a and b are two integers without any common factors.

A number such as π which *cannot* be expressed in the form $\frac{a}{b}$ is called **irrational**.

It is worth remembering that:

- all the square roots of numbers that are not square numbers are irrational. For example:

$$\sqrt{2}, \sqrt{3}, \sqrt{5}, \sqrt{6}, \sqrt{7}, \sqrt{8}, \sqrt{10}\ldots$$

  are all irrational numbers.

- the addition of a rational number to an irrational number will always create another irrational number. For example:

$$3 + \sqrt{2} \text{ and } \pi + 5$$

  are irrational numbers.

Surds

Some general rules of surds:

$$\sqrt{a} \times \sqrt{b} = \sqrt{ab} \qquad\qquad c\sqrt{a} \times d\sqrt{b} = cd\sqrt{ab}$$

$$\sqrt{a} \div \sqrt{b} = \sqrt{\frac{a}{b}} \qquad\qquad c\sqrt{a} \div d\sqrt{b} = \frac{c}{d}\sqrt{\frac{a}{b}}$$

Examples

- $\sqrt{12} = \sqrt{4} \times \sqrt{3} = 2\sqrt{3}$
- $\sqrt{3} \times \sqrt{3} = \sqrt{9} = 3$
- $\sqrt{3} \times \sqrt{5} = \sqrt{15}$
- $\sqrt{2} \times \sqrt{8} = \sqrt{16} = 4$
- $\sqrt{2} \times \sqrt{14} = \sqrt{28} = \sqrt{(4 \times 7)} = \sqrt{4} \times \sqrt{7} = 2\sqrt{7}$
- $\sqrt{6} \times \sqrt{15} = \sqrt{90} = \sqrt{9} \times \sqrt{10} = 3\sqrt{10}$

- $2\sqrt{5} \times 3\sqrt{3} = 6\sqrt{15}$
- $4\sqrt{6} \div 2\sqrt{3} = \dfrac{4}{2}\sqrt{\dfrac{6}{3}} = 2\sqrt{2}$

Indices

x^1 always has the value of x, whatever x may be.

x^0 always has the value of 1.

We use a negative index to show a fraction. For example:

$$x^{-a} = \frac{1}{x^a}$$

Examples

- $3^{-2} = \dfrac{1}{3^2}$
- $5^{-1} = \dfrac{1}{5}$
- $7x^{-2} = \dfrac{7}{x^2}$

Rules for multiplying and dividing numbers in index form

When we multiply numbers of the same variable in index form, we add the indices.

$$a^x \times a^y = a^{(x+y)}$$

For example:

$$10^2 \times 10^8 = 10^{10}$$
$$5^7 \times 5^6 = 5^{(7+6)} = 5^{13}$$
$$7^8 \times 7^{-3} = 7^5$$

There is a similar rule for dividing powers with the same base; we subtract the indices.

$$a^x \div a^y = a^{(x-y)}$$

For example:

$$a^5 \div a^3 = a^{(5-3)} = a^2$$
$$b^3 \div b^8 = b^{-5}$$

Powers of powers

$$(a^x)^y = a^{xy}$$

For example:

$$(t^2)^3 = t^{2\times3} = t^6$$
$$(t^{-2})^5 = t^{-10}$$
$$(t^3)^5 = t^{15}$$

Indices of the form $\frac{1}{n}$ (fractional indices)

$x^{\frac{1}{n}} = \sqrt[n]{x}$, that is, the nth root of x.

Examples

- $64^{\frac{1}{2}} = \sqrt{64} = 8$

- $125^{\frac{1}{3}} = \sqrt[3]{125} = 5$

- $10000^{\frac{1}{4}} = \sqrt[4]{10000} = 10$

- $25^{-\frac{1}{2}} = \frac{1}{\sqrt{25}} = \frac{1}{5}$

Indices of the form $\frac{a}{b}$

$x^{\frac{a}{b}} = (x^a)^{\frac{1}{b}} = \sqrt[b]{x^a}$ or $(\sqrt[b]{x})^a$

For example:

$16^{\frac{3}{4}} = \sqrt[4]{(16)^3} = 2^3 = 8$ or $(\sqrt[4]{16})^3$

Standard form

You can write large and small numbers using powers of 10.

A number written in standard form has the form

$a \times 10^n$

where $1 \leq a < 10$, and n is a whole number.

Some examples are:

- $82.1 = 8.21 \times 10 = 8.21 \times 10^1$
- $325 = 3.25 \times 100 = 3.25 \times 10^2$
- $4728 = 4.728 \times 1000 = 4.728 \times 10^3$
- $0.0034 = 3.4 \div 1000 = 3.4 \times 10^{-3}$
- $0.000000087 = 8.7 \div 100000000 = 8.7 \times 10^{-8}$

Notice that numbers *bigger than 10* are multiplied by a *positive* power of 10 and numbers *smaller than 1* are multiplied by a *negative* power of 10.

You need to be able to work with standard form numbers both with and without a calculator.

Example

- Without a calculator, work it out in two parts. For example:

  Calculate $4.5 \times 10^8 \times 1.2 \times 10^7$. Give your answer in standard form to 2 s.f.

  $$4.5 \times 10^8 \times 1.2 \times 10^7$$
  $$= 4.5 \times 1.2 \times 10^8 \times 10^7$$
  $$= 5.4 \times 10^{15}$$

- With a calculator, you can key in the numbers in standard form. For example:

  $$4.65 \times 10^5 \div 8.23 \times 10^7$$

  Key in:

  The answer may be displayed as:

  $$5.6500608^{-03} \text{ or } 5.6500608 \times 10^{-3}$$

  Check your own calculator.

  The answer is 5.65×10^{-3} (3 s.f.).

Test yourself

1 **(a)** Increase 5 kg by 17%.   **(b)** Decrease 5 m by 21%.

2 A baby dolphin increases its weight by 5% every day of its young life. What is the weight of a 5 day old baby dolphin which weighed 8.5 kg at birth?

3 Without using a calculator, estimate the value of the following:

  (a) 36.8×21.6 **(b)** $562 \div 6.7$

  (c) $\dfrac{134 \times 57.3}{7.5}$ **(d)** $8.3^2 + 9.7^2$

4 Divide £325 in the ratio 2:3.

5 If £23 is 20% of an amount, what is 100% of the amount?

6 Work out:

  (a) $\dfrac{2}{3} + \dfrac{3}{5}$ **(b)** $\dfrac{7}{8} - \dfrac{4}{5}$

(c) $\dfrac{3}{4} \times \dfrac{8}{9}$ **(d)** $\dfrac{5}{6} \div \dfrac{7}{12}$

7 Write down the limits of accuracy for the following measurements:

(a) 4 cm **(b)** 1.95 m **(c)** 7.23 kg

8 Evaluate the following:

(a) $81^{\frac{1}{2}}$ **(b)** $27^{\frac{1}{3}}$ **(c)** $512^{\frac{1}{3}}$ **(d)** $625^{\frac{3}{4}}$ **(e)** $100\,000^{\frac{4}{5}}$

9 Simplify the following surds:

(a) $5\sqrt{12} \times 3\sqrt{3}$ *90* **(b)** $4\sqrt{10} \times 3\sqrt{5}$ **(c)** $4\sqrt{18} \times 3\sqrt{14}$

60   9

10 Write down:

(a) two irrational numbers between 5 and 6

(b) two irrational numbers between 50 and 60

11 Write the following numbers in standard form:

(a) 5684 **(b)** 5 million **(c)** 38 200 **(d)** 0.0095

12 Calculate the following without a calculator, leaving your answer in standard form:

(a) $3.4 \times 10^3 \times 2.5 \times 10^9$ **(b)** $8.4 \times 10^4 \div 1.2 \times 10^9$

13 Calculate the following using a calculator. Write your answer in standard form to 3 significant figures:

(a) 9 million $\div 7.32 \times 10^{17}$ **(b)** $5.67 \times 10^{-6} \times 8.39 \times 10^{-7}$

Bobba

Practice questions

1 Do not use a calculator when answering this question. All working must be shown.

Obtain an *estimate* for $\dfrac{46.7 \times 0.44}{6.23}$

2 In a right-angled triangle ABC, the height is 8.6 cm and the base is 6.3 cm, both measurements to the nearest millimetre. Calculate the upper and lower limits of accuracy of the area of the triangle.

3 Three people are in a lottery syndicate. Malcolm has 4 sets of numbers, Janet has 6 sets of numbers and Neil has 18 sets of numbers. They win a big prize of £3 456 576 which they share out in the ratio of the number of sets of numbers they have. How much does each person in the syndicate receive?

4 I am told that, travelling at an average speed of 45 miles per hour (2 s.f.), it will take me 8 minutes (1 s.f.) to travel from school to the library.

(a) Use this information to calculate:
  (i) the upper limits of accuracy of the distance from school to the library
  (ii) the lower limits of accuracy of the distance from school to the library

(b) From part **(a)**, what would you say is the sensible figure to give as the distance between the school and the library?

5 Mark has two drill bits marked $\frac{7}{8}$ inches and $\frac{5}{16}$ inches. What is the drill size halfway between these two values?

6 Carl puts £5000 in a building society account. Compound interest at 4.8% is added at the end of each year.

(a) Calculate the total amount of money in Carl's savings account at the end of 3 years. Give your answer to the nearest pound.

Mandy also puts a sum of money in a building society savings account. Compound interest at 5% is added at the end of each year.

(b) Work out the single number by which Mandy has to multiply her sum of money to find the total amount she will have after 3 years.

7 (a) Evaluate $64^{\frac{1}{3}}$.

(b) Write $16^{\frac{1}{2}} \times 8^{-3}$ as a power of 2.

(c) Given that $125^y = 5$, find the value of y.

8 (a) Which of the following numbers are rational?
  (i) $5 + \sqrt{3}$ (ii) π^2 (iii) $5^3 + 5^0 + 5^{-3}$

(b) When A and B are two different irrational numbers, $A \times B$ can be rational. Write down two different examples to show this.

9 (a) Show clearly that $(\sqrt{3} + \sqrt{12})^2 = 27$.

(b) Given that $x = \sqrt{6}$, $y = \sqrt{18}$ and $z = \sqrt{27}$, evaluate the following:

    (i) $\dfrac{y}{x}$ (ii) xyz

10 Simplify $\sqrt{18} + \sqrt{72}$.

11 **(a)** The approximate population of the UK is given in standard form as 5.4×10^7. Write this as an ordinary number.

    (b) If the total population of the UK spent two and a half billion pounds on food in 1 week, how much on average did each person spend?

12 The world harvest of garlic is 20 000 tonnes every day.

    (a) How much garlic is harvested in 1 year? (Assume a year is 365 days.) Give your answer in standard form.

France produces 5.29×10^4 tonnes of garlic in a year.

    (b) What percentage of the world total is produced by France?

Algebra

Solving linear equations

Example

Solve the following equation:

$$5(2x - 9) = 14$$

If we multiply out the brackets first we get:

$$10x - 45 = 14$$
$$10x = 14 + 45 = 59$$
$$x = \frac{59}{10} = 5.9$$
$$x = 5.9$$

Simultaneous equations

Example 1

Solve the simultaneous equations:

$$6x + y = 15$$
$$4x + y = 11$$

Since both equations contain a y term, we can subtract one equation from the other to give

$$2x = 4$$

which solves to give

$$x = 2$$

We can now substitute $x = 2$ into one of the first equations (usually the one with the smallest numbers). So substitute into:

$$4x + y = 11$$

to give

$$8 + y = 11$$

which gives

$$y = 11 - 8$$
$$y = 3$$

We test the solution by substituting back into the other first equation. Substituting $x = 2$ and $y = 3$ into $6x + y$ gives $12 + 3 = 15$, which is correct. So we can confidently say our solution is $x = 2$ and $y = 3$.

Example 2

Solve the simultaneous equations:

$$4x - 2y = 12$$
$$2x + 2y = 18$$

Since both equations contain $2y$ but one has a $+$ and the other a $-$ then we can add the two equations to give

$$6x = 30$$
$$x = 5$$

Substitute $x = 5$ into, say, the bottom equation to get

$$2 \times 5 + 2y = 18$$
$$10 + 2y = 18$$
$$2y = 18 - 10 = 8$$
$$y = 4$$

The solution of $x = 5$ and $y = 4$ can be checked in the top equation to give $4 \times 5 - 2 \times 4 = 20 - 8 = 12$, which is correct. So our solution is $x = 5$ and $y = 4$.

Example 3

Solve the simultaneous equations:

$$4x + 2y = 32$$
$$3x - y = 19$$

Here we do not have any equal terms so we have to start creating equal terms because that is the only way we can solve simultaneous equations!

We can see that by multiplying *all* of the second equation by 2 we get

$$(3x - y = 19) \times 2 \Rightarrow 6x - 2y = 38$$

Our pair of equations is now:

$$4x + 2y = 32$$
$$6x - 2y = 38$$

and we can solve these as we did in Example 2.

Example 4

Solve the simultaneous equations:

$$5x + 4y = 22$$
$$2x + 3y = 16$$

Notice that we cannot simply multiply one equation by anything to give us equal terms. So we have to multiply *both* equations.

The choice is now up to us. We can either make the x values the same or the y values the same. Sometimes there is an obvious choice; sometimes it does not matter. In this example it does not matter since there is no great advantage in choosing either.

Let us make the x values equal.

We will have to multiply the first equation through by 2 and multiply the second equation through by 5. This gives:

$$(5x + 4y = 22) \times 2 \Rightarrow 10x + 8y = 44$$

and

$$(2x + 3y = 16) \times 5 \Rightarrow 10x + 15y = 80$$

We can now solve these in the same way as we did in Example 1.

Problems solved by simultaneous equations

You may be given a problem that involves a simultaneous equation. For example:

Two teas and five buns cost £1.04, and three teas and two buns cost £1.01. How much is tea and a bun?

In this type of question, you replace the words tea and bun with t and b so the equations to solve become:

$$2t + 5b = 104$$
$$3t + 2b = 101$$

and solve as before.

Solving linear inequalities

Inequalities are treated similarly to normal equations. The difference is that they have an inequality sign instead of an equals sign.

We use the same rules to solve linear inequalities as we do for linear equations.

Example 1

Solve the inequality:

$$\frac{5x + 7}{3} < 14$$

Multiplying both sides by 3 we get

$$5x + 7 < 14 \times 3$$
$$5x + 7 < 42$$

Then subtracting 7 from both sides gives

$$5x < 42 - 7$$
$$5x < 35$$

Finally, dividing both sides by 5 gives

$$x < 35 \div 5$$
$$x < 7$$

Example 2

Solve the inequality:

$$1 < 5x + 3 \le 17$$

We need to treat each side separately:

$1 < 5x + 3$	$5x + 3 \le 17$
$1 - 3 < 5x$	$5x \le 17 - 3$
$-2 < 5x$	$5x \le 14$
$\dfrac{-2}{5} < x$	$x \le \dfrac{14}{5}$
$-0.4 < x$	$x \le 2.8$

Hence

$$-0.4 < x \le 2.8$$

Inequalities involving x^2

Consider $x^2 < 16$.

If we had the equation $x^2 = 16$, we know the two solutions would be 4 and -4.

When we look at $x = 4$, we can see that $x < 4$ fits the solution well. But $x = -4$ doesn't because $x < -4$ just doesn't work. (For example, $-5 < -4$ but $-5^2 = 25$, which is not less than 16.)

In fact, the solution connected with $x = -4$ needs the inequality sign changed around to give the solution $x > -4$, or $-4 < x$.

The solution can be shown on a number line:

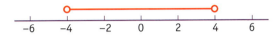

i.e. $-4 < x < 4$.

Solve the inequality:

$$x^2 > 25$$

The solution to $x^2 > 25$ will be $x > 5$ and $x < -5$.

On a number line this is shown as:

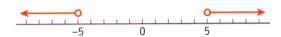

Notice the difference between the two types, $x^2 < a^2$ and $x^2 > a^2$, as shown on the number lines.

Expansion

This is where we multiply everything inside the bracket by what is outside the bracket. Some examples of expansion are:

- $2(3t + 4) = 6t + 8$
- $k(p + 6) = kp + 6k$
- $3t(t + 5) = 3t^2 + 15t$
- $p(p^2 - 5x) = p^3 - 5px$
- $4x^2(2x + 7) = 8x^3 + 28x^2$
- $-3t(2 - 4t) = -6t + 12t^2$
- $4t(1 + 3t - q) = 4t + 12t^2 - 4qt$

Simplification

This is tidying up, often after an expansion. It involves adding or subtracting like terms (those that are the same power of the same variable).

Example 1

$3(5 + p) + 2(4 + 3p)$
$= 15 + 3p + 8 + 6p = 23 + 9p$

algebra

Example 2

$3w(4w + 1) - w(3w - 7)$
$= 12w^2 + 3w - 3w^2 + 7w = 9w^2 + 10w$

Transformation of formulae

Just as the terms in equations can be moved around, so can formulae. Look at the following examples to see how this is done.

Example 1

From the formula $T = 4m - 1$, make m the subject:

$T + 1 = 4m$ (moving the –1 away from the $4m$)

$\dfrac{T + 1}{4} = m$ (moving the 4 away from the m)

Hence the transformed formula becomes $m = \dfrac{T + 1}{4}$

Example 2

From the formula $Q = 7(2t + 3)$, express t in terms of Q (this is another common way of asking you to make t the subject).

$Q = 14t + 21$ (expanding the bracket)

$Q - 21 = 14t$ (moving the 21 away from the $14t$)

$\dfrac{Q - 21}{14} = t$ (moving the 14 away from the t)

Hence the transformed formula becomes $t = \dfrac{Q - 21}{14}$

Example 3

From the formula $T = r^2 - p$
(i) make r the subject
(ii) make p the subject

(i) Make r the subject:

$T + p = r^2$ (moving the p away from the r^2)

$\sqrt{(T + p)} = r$ (taking the square root of both sides)

Hence the transformed formula becomes $r = \sqrt{(T + p)}$

(ii) Make p the subject:

$T + p = r^2$ (move p to make it positive)

$p = r^2 - T$ (move the T away from the p)

Hence the transformed formula becomes $p = r^2 - T$

Factorisation

This is the opposite of expansion. It is, in effect, putting the brackets back into the expression. A common factor is put outside the brackets, as shown in the following examples.

Examples

- $4t + 6m = 2(2t + 3m)$
- $6mt + 9pt = 3t(2m + 3p)$
- $5p^2 - 25p = 5p(p - 5)$
- $12px + 4tx - 8mx = 4x(3p + t - 2m)$

Quadratic expansion

When we have an expression such as $(5y + 2)(2y - 3)$ then it can be expanded. The result is a quadratic expression. A quadratic expression contains a term with a power of 2 (e.g. x^2).

Multiplying out such pairs of brackets is called quadratic expansion.

The rule for expanding expressions such as $(x + 4)(3x - 3)$ is similar to expanding single brackets: 'multiply everything in one bracket by everything in the other bracket'.

Example 1

$$\begin{aligned}(y + 6)(y - 3) &= y(y - 3) + 6(y - 3) \\ &= y^2 - 3y + 6y - 18 \\ &= y^2 + 3y - 18\end{aligned}$$

Example 2

$$\begin{aligned}(t - 2)(t + 1) &= t(t + 1) - 2(t + 1) \\ &= t^2 + t - 2t - 2 \\ &= t^2 - t - 2\end{aligned}$$

Example 3

$$\begin{aligned}(q - 4)(q - 2) &= q(q - 2) - 4(q - 2) \\ &= q^2 - 2q - 4q + 8 \\ &= q^2 - 6q + 8\end{aligned}$$

Higher Tier

Example 4

$$(3t + 2)(2t - 1) = 3t(2t - 1) + 2(2t - 1)$$
$$= 6t^2 - 3t + 4t - 2$$
$$= 6t^2 + t - 2$$

Expanding squares

Example

$$(2x - 5)^2 = (2x - 5)(2x - 5) = 2x(2x - 5) - 5(2x - 5)$$
$$= 4x^2 - 10x - 10x + 25$$
$$= 4x^2 - 20x + 25$$

Quadratic factorisation

This is the reverse of quadratic expansion. It is putting the expansion back into brackets.

Example 1

Factorise $x^2 + 9x + 20$

- We note that both brackets start with an x: $(x \quad)(x \quad)$
- The last sign is positive; hence both bracket signs will be the same as the first sign in the expansion, which is plus:
  $(x + \quad)(x + \quad)$
- We note that the end bracket numbers multiply together to give 20
  i.e. 1 and 20 or 2 and 10 or 4 and 5
- The signs in the brackets are the same, so the end bracket numbers add up to 9; hence they must be 4 and 5
- So the brackets become: $(x + 4)(x + 5)$

Example 2

Factorise $x^2 - 7x + 10$

- We note that both brackets start with an x: $(x \quad)(x \quad)$
- The last sign is positive, so both bracket signs will be the same as the first sign in the expansion, which is minus:
  $(x - \quad)(x - \quad)$
- We note that the end bracket numbers multiply together to give 10
  i.e. 1 and 10 or 2 and 5
- The signs in the brackets are the same, so the end bracket numbers add up to 7; hence they must be 2 and 5
- So the brackets become: $(x - 2)(x - 5)$

Example 3

Factorise $x^2 + 3x - 4$

- We note that both brackets start with an x: $(x \quad)(x \quad)$
- The last sign is negative, which means that the signs in the brackets are different:

$$(x + \quad)(x - \quad)$$

- We note that the end bracket numbers multiply together to give 4
  i.e. 2 and 2 or 1 and 4
- The signs in the brackets are different, so the end bracket numbers have a difference of 3; hence they must be 1 and 4
- Since the middle term of the expansion is positive, the largest end number must be with the positive sign.
- So the brackets become: $(x + 4)(x - 1)$

Difference of two squares

Factorise $x^2 - n^2$.

This factorises to:

$$(x + n)(x - n)$$

- $x^2 - 25 = (x + 5)(x - 5)$
- $x^2 - y^2 = (x + y)(x - y)$
- $x^2 - 9m^2 = (x + 3m)(x - 3m)$

Solving quadratic equations as $ax^2 + bx + c = 0$

If the exam question does *not* ask for the answer to be given to 1 or 2 decimal places, then the quadratic equation will almost certainly factorise as shown in the following example:

Example 1

Solve $x^2 + 2x - 15 = 0$

This factorises into:

$$(x - 3)(x + 5) = 0$$

The only way this expression can equal 0 is if one of the brackets is worth 0.

Hence either $(x - 3) = 0$ or $(x + 5) = 0$

Hence $x - 3 = 0$ or $x + 5 = 0$

Hence $x = 3$ or $x = -5$

The solution then is $x = 3$ and $x = -5$

algebra

If the question asks you to solve a quadratic equation to 1 or more decimal places, then you will undoubtedly need to use the formula.

Formula for solving quadratic equations

The following formula can be used to solve any quadratic equation (or to tell you that there is no solution).

For a quadratic equation $ax^2 + bx + c = 0$

$$x = \frac{-b \pm \sqrt{(b^2 - 4ac)}}{2a}$$

$\pm$ is used because a square root has a positive and a negative solution.

Example

Solve the equation $3x^2 - 8x + 2 = 0$, correct to 2 decimal places.

$$\text{Use } x = \frac{-b \pm \sqrt{(b^2 - 4ac)}}{2a} \text{ where } a = 3, b = -8 \text{ and } c = 2$$

$$= \frac{8 \pm \sqrt{(64 - 4 \times 3 \times 2)}}{6}$$

$$= \frac{8 \pm \sqrt{(40)}}{6} = \frac{8 + \sqrt{(40)}}{6} \text{ and } \frac{8 - \sqrt{(40)}}{6}$$

$$= 2.39 \text{ or } 0.28$$

It is useful to work out $\sqrt{(b^2 - 4ac)}$ on your calculator first and put the value into the calculator memory to use twice in the solution.

Exam tip: If the question asks you to solve a quadratic equation to 1 or 2 decimal places, then you can be sure it can only be solved by the formula.

Direct proportionality

If one variable is directly proportional to another variable, then as one gets bigger, so the other also gets bigger, in the same proportion.

The symbol we use for variation or proportion is $\propto$.

So the statement 'weight is directly proportional to height' can be written mathematically as:

weight $\propto$ height

and if this is true, then we can say that

weight $= k \times$ height

where k is the constant of proportionality.

Example 1

The price of a model is directly proportional to the time spent making it.
A model that takes 4 hours to make, costs £22. Find:
(i) the cost of a model that takes $7\frac{1}{2}$ hours to make
(ii) the length of time it takes to make a model costing £35

First, we can say that price $\propto t$, hence price $= kt$.

Since price $=$ £22 when time $=$ 4 hours then

$$22 = 4k$$
$$\frac{22}{4} = k$$

Hence $k = 5.5$

So our correct formula is price $= 5.5t$

(i) When time $= 7\frac{1}{2}$ hours:

$$\text{price} = 5.5 \times 7.5 = 41.25$$
$$\text{price} = £41.25$$

(ii) When cost $=$ £35:

$$35 = 5.5 \times \text{time}$$
$$\frac{35}{5.5} = \text{time}$$

Hence time $= 6.36$ hours (or 6 hours 22 minutes)

Example 2

The weight of a rubber ball is directly proportional to the square of its radius. The weight of a rubber ball with a radius of 5 cm is 535 grams. Find:
(a) the weight of a rubber ball of radius 3 cm
(b) the radius of a rubber ball weighing 1 kg

Let $W =$ weight, $r =$ radius

The first sentence tells us that

$$W \propto r^2$$

Hence $W = kr^2$

Weight $= 535$ grams when radius $= 5$ cm, so

$$535 = 25k$$
$$\frac{535}{25} = k = 21.4$$

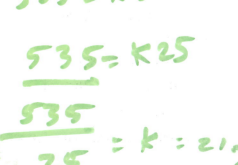

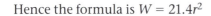

Hence the formula is $W = 21.4r^2$

(a) When radius = 3 cm

$$W = 21.4 \times 3^2 = 192.6 \text{ grams}$$

Rounding off gives weight = 193 grams

Notice that, because the question did not specify the number of significant figures, we have given the answer to the same accuracy as the values given in the question (3 s.f.).

(b) When weight = 1 kg = 1000 grams

$$1000 = 21.4r^2$$

Hence $\dfrac{1000}{21.4} = r^2$

so $r^2 = 46.729$

Hence $r = \sqrt{46.729} = 6.835$

This gives us a radius of 6.84 cm (to 3 s.f.).

Inverse variation or inverse proportion

This is where there is a dividing ratio between two variables, so that as one variable increases, the other decreases and vice versa. For example, as you travel faster you cover distance in less time. So there is an inverse variation between speed and time. We say speed is inversely proportional to time:

$$S \propto \frac{1}{T}$$

That is, if A is inversely proportional to B then

$$A \propto \frac{1}{B}$$

and so

$$A = k \times \frac{1}{B} = \frac{k}{B}$$

where k is the constant of proportionality.

Example

P is inversely proportional to d^2. If $P = 4.5$ when $d = 4$, find:
(i) P when $d = 8$
(ii) d when $P = 50$

The first statement tells us that $P \propto \dfrac{1}{d^2}$

Hence $P = \dfrac{k}{d^2}$

So when $P = 4.5$ and $d = 4$ then $4.5 = \dfrac{k}{16}$

Hence $4.5 \times 16 = k = 72$

Hence the formula is $P = \dfrac{72}{d^2}$

(i) When $d = 8$

$$P = \frac{72}{64} = 1.125$$

(ii) When $P = 50$

$$50 = \frac{72}{d^2}$$

so $50d^2 = 72$

so $d^2 = \dfrac{72}{50} = 1.44$

so $d = \sqrt{1.44} = 1.2$

General or *n*th term of a linear sequence

A linear sequence has the *same difference* between each pair of consecutive terms. For example:

- 2, 5, 8, 11, 14...   difference of 3
- 5, 7, 9, 11, 13...   difference of 2

The general term or *n*th term of a linear sequence is always of the form '*an* + *b*', where the coefficient of *n* is the difference between each term and *b* is the difference between the first term and *a*.

For the sequence 2, 5, 8, 11, 14... find:
(a) the *n*th term
(b) the 50th term
(c) the first term that is greater than 1000

The difference between consecutive terms is 3. So the first part of the *n*th term is 3*n*.

algebra

Subtract the difference 3 from the first term 2 to give $2 - 3 = -1$

(a) So the nth term is given by $3n - 1$

(b) The 50th term is found by substituting $n = 50$ into the nth term, $3n - 1$

So the 50th term $= 3 \times 50 - 1 = 150 - 1 = 149$

(c) The first term that is greater than 1000 is given by:

$$3n - 1 > 1000$$
$$3n > 1000 + 1$$
$$n > \frac{1001}{3}$$
$$n > 333.667$$

So the first term (which has to be a whole number) over 1000 is the 334th.

General or nth term of a quadratic sequence

These sequences will be based on n^2, so you do need to recognise the pattern 1, 4, 9, 16, 25...

The differences between the consecutive terms of this pattern are the odd numbers 3, 5, 7, 9...

So if you find that the differences form an odd number sequence, you know the pattern is based on n^2.

Example

Find the nth term in the sequence

2, 5, 10, 17, 26...

The differences are the odd numbers 3, 5, 7, 9... so we know the rule is based on n^2.

We look for a link with the square numbers. We do this by subtracting from each term the corresponding square number.

2	5	10	17	26
-1	-4	-9	-16	-25
1	1	1	1	1

Clearly, the link is $+1$, so the nth term is $n^2 + 1$.

Test yourself

1 Solve the following pairs of simultaneous equations:

(a) $5x + y = 0$
$3x - 2y = 13$

(b) $7x + 3y = 18$
$x + y = 4$

2 The cost of hiring a hall is directly proportional to the square root of the number of people attending the function. When 100 people attend, the hall costs £75 to hire.

(a) How much will it cost to hire the hall for a party of 200?

(b) How many people attended a function when the hire cost was £97.50 for the hall?

3 Expand the following and simplify:

(a) $(2x - 3)(4x + 1)$ **(b)** $(3x + 5)(x - 3)$ **(c)** $p(2m + t) - t(3m - p)$

4 Solve the following equations:

(a) $2x - 3 = 11$ **(b)** $3 - 5x = 8$ **(c)** $4(2x - 3) = 7$

5 Make y the subject of the following formulae:

(a) $x = 2(y - 1)$ **(b)** $x = y(b + 7)$ **(c)** $t = \dfrac{5y + p}{7}$

6 Factorise the following:

(a) $3t + 7t^2$ **(b)** $2m^3 - 6m^2$ **(c)** $6mp^3 + 9m^2pt$

(d) $x^2 - 7x + 12$ **(e)** $x^2 - 25$ **(f)** $2x^2 - x - 15$

7 (a) Solve the quadratic equation $x^2 + 7x + 12 = 0$.

(b) Solve the quadratic equation $5x^2 + 6x - 2 = 0$ to 2 decimal places.

8 Solve the following inequalities:

(a) $5x > 32$ **(b)** $4t < 5t - 8$ **(c)** $x^2 < 36$ **(d)** $-2 \leq 5x + 3 < 4$

9 State the nth term for the following sequences:

(a) 2, 7, 12, 17... **(b)** 3, 6, 11, 18, 27... **(c)** $\dfrac{1}{5}, \dfrac{4}{9}, \dfrac{7}{13}, \dfrac{10}{17}...$

Practice questions

1 Given that $w = 7.5$, $x = 0.35$ and $y = -5$, calculate:

(a) $w(x + y^2)$

(b) $\dfrac{(w^2 - x^2)}{y}$

2 n is a non-zero integer. $\dfrac{58}{n^2} > 5$. List all the possible values of n.

3 Solve the equations:

(a) $5x - 3 = 40$

(b) $5(4x + 5) = 14$

(c) $\dfrac{3x + 6}{x} = 20$

4 (a) Multiply out and simplify $(x + a)(x - a)$.

(b) Show how you can use part **(a)** to find the exact value of $978.5^2 - 21.5^2$.

5 Solve the simultaneous equations:

$4x + 2y = 11$

$5x - 3y = 22$

6 (a) (i) Solve the inequality $4n - 7 < 52 + 9n$.

(ii) Write down the least whole number value of n that satisfies this inequality.

(b) Solve the inequality $x^2 < 1.44$.

7 The cost of producing a gold ring is directly proportional to the square of the radius of the ring. A gold ring of radius 7 mm costs £10.55 to produce.

(a) What is the cost of producing a gold ring with a radius of 9 mm?

(b) What is the radius of a gold ring that costs £26.05 to produce?

8 (a) Expand and simplify $(4x + 5)(3x - 2)$.

(b) Factorise completely $8x - 12x^2$.

(c) From the formula $T = \dfrac{\sqrt{(9 - 5M)}}{7}$ make M the subject.

(d) Express $\dfrac{x^{-3}y}{xy^3}$ in the form $x^a y^b$, stating clearly the values of a and b.

9 (a) (i) Write the equation $x^2 + 6x + 11$ in the form $(x + a)^2 + b$, where a and b are integers.

(ii) Hence, write down the minimum value of $x^2 + 6x + 11$.

(b) Solve the equation $4x^2 + 6x - 11 = 0$. Give your answer to 2 decimal places.

10 (a) A sequence is given by $\dfrac{2}{3}, \dfrac{4}{5}, \dfrac{6}{7}, \dfrac{8}{9}, \ldots$

   (i) write down the 12th term of this sequence
   (ii) write down the nth term of this sequence

(b) Each term of another sequence is the reciprocal of each term in the sequence of part **(a)**. Write down the first three terms of that sequence.

Graphs

Gradients

The gradient of a straight line is a measure of its slope.

The gradient of the line shown here can be measured by drawing a right-angled triangle (as big as possible), having the line as the hypotenuse (sloping side).

The gradient is then found by the following rule:

$$\text{gradient} = \frac{\text{vertical distance}}{\text{horizontal distance}}$$

Travel graphs

Distance–time graphs

On a distance–time graph, the gradient tells us the average speed between any two times, and the speed at any particular instant in time.

The diagram below shows a car journey over a 4-hour period.

The graph is a curve.

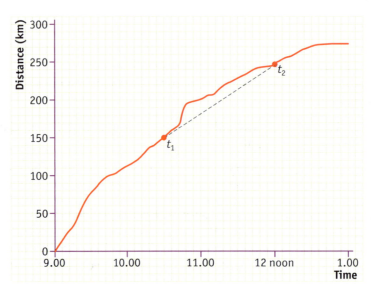

Average speed

The average speed between time t_1 and time t_2 is found by calculating the gradient of the chord (straight line) joining the points on the curve corresponding to those times.

For example, to find the average speed between 10.30 a.m. and 12 noon, we draw the chord t_1t_2.

Average speed = gradient of $t_1t_2 = \dfrac{100}{1.5} = 67$ km per hour.

Instantaneous speed

The instantaneous speed is constantly changing along a curved line. However, at any given time, t, the instantaneous speed can be estimated by the gradient of the tangent to the distance–time curve at that exact time, t.

To draw a tangent, you need to put a straight edge (ruler) on the line so that the point at which you wish to draw the tangent is the only point of the curve seen in that part of the graph.

To estimate the speed, draw the tangent of the curve at that point and calculate the gradient.

For example, to find the instantaneous speed at 10.48 a.m., draw the tangent to the curve at $t = 10.48$ a.m.

This is shown as $CD = \dfrac{100}{1} = 100$ km per hour.

Note that when you have a negative gradient on a distance–time graph, this gives the speed on the return journey.

Velocity–time graphs

When we looked at speed we ignored the sign. If we were to take notice of the sign of the gradient, then it would tell us the direction of travel. Once we take particular note of the direction of travel, then we are looking at velocity instead of speed.

Velocity is found from the gradient of a distance–time graph at time t, taking the sign of the gradient into account.

When the velocity of a moving object is plotted against time, two other pieces of information can be found from the graph.

- The gradient of a velocity–time graph at any time t gives the acceleration of the object at that time. Acceleration is the rate of change of velocity. The units of acceleration are m/s² or km/h². If the gradient of a velocity–time graph is negative, then the speed is decreasing and the object is decelerating.

- The area under the graph between P_1 and P_2 (that is, the area bounded by the graph, the horizontal axis and the lines $t = t_1$ and $t = t_2$) gives the distance travelled between t_1 and t_2.

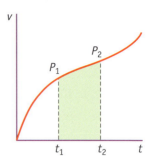

Example

The velocity of a particle is shown on the velocity–time graph below.

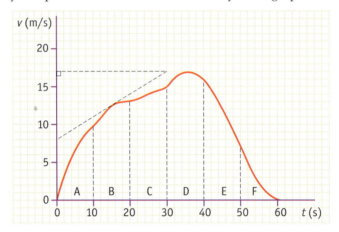

Calculate:

(i) the acceleration after 15 s

(ii) the total distance travelled over the 60 s

(i) The acceleration after 15 s is found by drawing the tangent at $t = 15$ and finding the gradient.

Using the triangle in the diagram, the gradient is $\dfrac{9}{30} = 0.3 \, \text{m/s}^2$

(ii) The total distance is given by the area under the curve. This can be calculated by dividing the area into two triangles and four trapeziums. (The method for finding the area of a trapezium is described on page 51.)

$$\tfrac{1}{2}(10 \times 10) + \left(10 \times \frac{(10 + 13)}{2}\right) + \left(10 \times \frac{(13 + 15)}{2}\right) +$$

[triangle A]　　　　[trapezium B]　　　　[trapezium C]

$$\left(10 \times \frac{(15 + 16)}{2}\right) + \left(10 \times \frac{(16 + 7)}{2}\right) + \tfrac{1}{2}(10 \times 7)$$

[trapezium D]　　　　[trapezium E]　　　　[triangle F]

$$= 50 + 115 + 140 + 155 + 115 + 35 = 610 \, \text{m}$$

You could also estimate the area by counting the grid squares. This is a valid method if carried out accurately.

Graphs from linear equations

These will always be a straight line.

You only need two points to be sure of your line, but plotting a third point is a useful check that the line is correct.

Example

Draw the graph of $y = 2x - 3$.

Choose different values of x and find the corresponding value of y when you substitute the x values into $y = 2x - 3$.

x	0	2	4
y	-3	1	5

Plot these points on a grid and join them up to see the straight line.

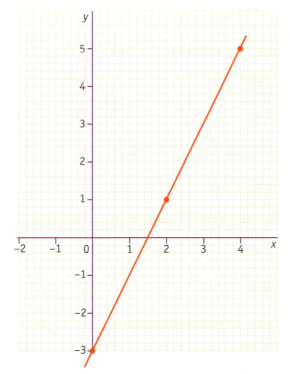

Every linear equation can be written in the form $y = mx + c$. If this is plotted as a graph, you will find that m = the gradient of the line and c = the point where the line cuts the y-axis.

Another special type of straight line is one given an equation of the form $ax + by = c$.

These types of equation can be drawn very easily without much working at all. The following is known as the 'cover and draw' method.

Example

Draw the graph of the equation $3x + 4y = 18$.

Cover up the $3x$ to leave $4y = 18$ which gives $y = 4.5$ at $x = 0$

Cover up the $4y$ to leave $3x = 18$ which gives $x = 6$ at $y = 0$

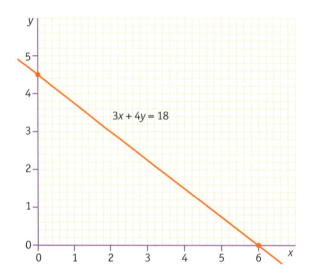

Graphs from quadratic equations

A quadratic equation can always be written in the form:

$$y = ax^2 + bx + c$$

The c is just like in the linear equation. It tells where the curve cuts through the y-axis.

Any graph from a quadratic equation will look like:

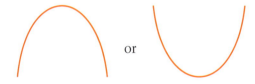

or

Because it is a curve, you will need to plot several points in order to find its shape.

You will always be given a table of values or at least the range of x values to use to find the points to plot.

There are two important points to remember when drawing quadratic graphs:

- Draw smooth curves.
- Curves have rounded bottoms!

Draw the graph of $y = x^2 + 3x - 1$.

x	−4	−3	−2	−1	0	1	2
x^2	16	9	4	1	0	1	4
3x	−12	−9	−6	−3	0	3	6
−1	−1	−1	−1	−1	−1	−1	−1
y	3	−1	−3	−3	−1	3	9

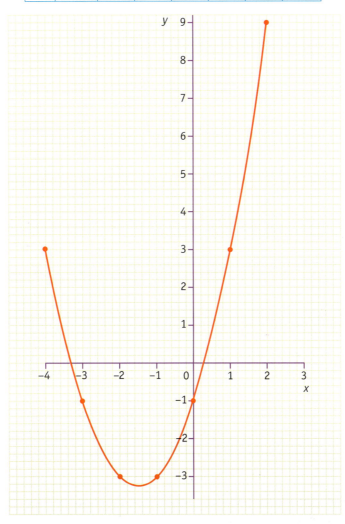

Graphs from reciprocal equations

A reciprocal equation is of the form $y = \dfrac{A}{x}$.

graphs

The shape is always like this.

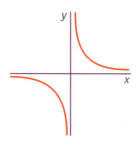

It is impossible to find values for $x = 0$ and $y = 0$ because the closer x gets to 0 the larger $\dfrac{A}{x}$ becomes.

Because of its complex shape, you will usually be given a table of values to use to draw the graph.

Example

Draw the graph of $y = \dfrac{6}{x}$ for $-7 < x < 7$.

We can draw up a table of values:

x	−6	−3	−2	−1	1	2	3	6
y	−1	−2	−3	−6	6	3	2	1

Plot the points and you get the following shape:

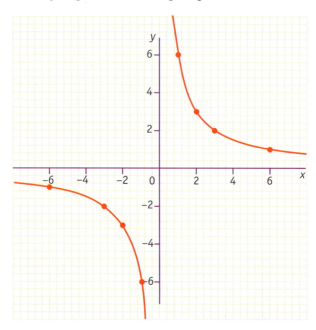

Solving simultaneous equations graphically

Linear simultaneous equations

Example 1

By drawing graphs, find the solution of the simultaneous equations:

(i) $4x + y = 8$

(ii) $y = 3x - 2$

(i) The first graph can be drawn using the 'cover and draw' method. It crosses the x-axis at $(2, 0)$ and the y-axis at $(0, 8)$.

(ii) The second graph can be drawn by finding some points by substituting x values, say $(0, -2)$, $(1, 1)$ and $(3, 7)$.

Hence we can see that the point where the graphs intersect is $(1.4, 2.3)$.

The solution to the simultaneous equations is $x = 1.4$, $y = 2.3$.

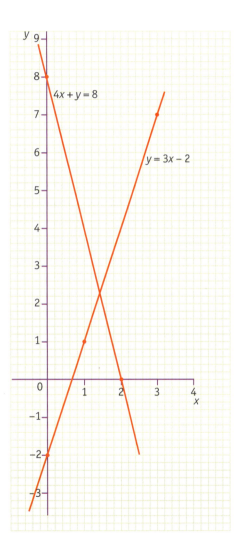

Non-linear simultaneous equations

You can solve a pair of non-linear simultaneous equations by drawing the graphs and finding the points of intersection.

graphs

Solving awkward looking equations

You can also solve equations such as $x^2 + 3x = \dfrac{1}{x + 1}$ by drawing the graph of each side.

For example, draw $y = x^2 + 3x$ and $y = \dfrac{1}{x + 1}$ and then find the points of intersection.

Trigonometric graphs

You need to be familiar with the basic shapes of the trigonometric graphs.

$y = \sin x$

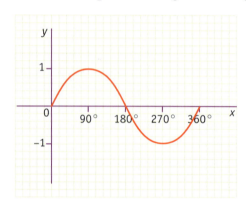

$y = \cos x$

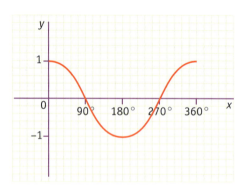

$y = \tan x$

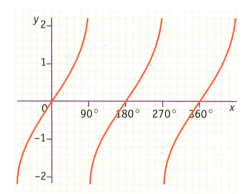

Transformations of the graph of $y = f(x)$

This graph represents any function $y = f(x)$.

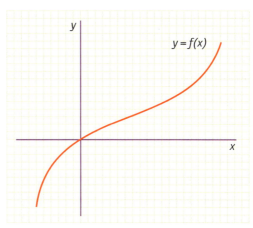

Rule 1

The graph of $y = f(x) + a$ is a translation of the graph of $y = f(x)$ by a vector $\begin{pmatrix} 0 \\ a \end{pmatrix}$

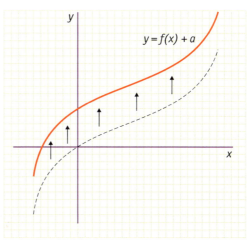

Rule 2

The graph of $y = f(x + a)$ is a translation of the graph of $y = f(x)$ by a vector $\begin{pmatrix} -a \\ 0 \end{pmatrix}$

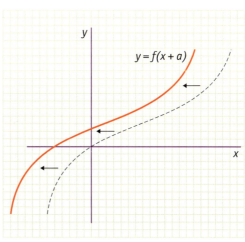

graphs

Rule 3

The graph of $y = kf(x)$ will be a stretch of the graph $y = f(x)$ by a scale factor of k in the y-direction.

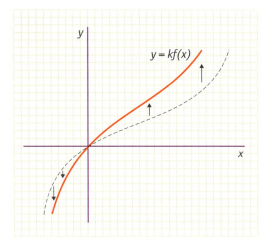

Rule 4

The graph of $y = f(tx)$ will be a stretch of the graph $y = f(x)$ by a scale factor of $1/t$ in the x-direction.

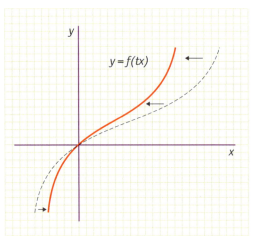

Rule 5

The graph of $y = -f(x)$ will be a reflection of the graph $y = f(x)$ in the x-axis.

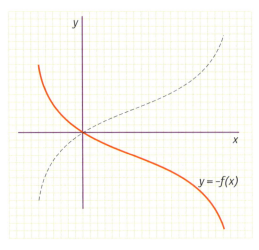

Rule 6

The graph of $y = f(-x)$ will be a reflection of the graph $y = f(x)$ in the y-axis.

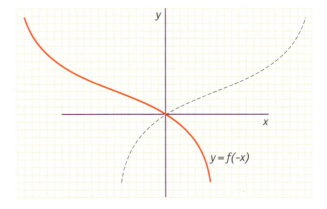

Finding regions of graphical inequalities

A two-dimensional inequality is a region that will be on one side or the other of a line.

You will recognise inequalities by the fact that they look like an equation but, instead of the equals sign, they have an inequality sign:

- $<$ less than
- $>$ greater than
- $\leq$ less than or equal to
- $\geq$ greater than or equal to

The method for drawing an inequality is to draw the 'boundary line', which is found by replacing the inequality with an equals sign.

After the boundary is drawn, the appropriate side of the line is shaded. This is found by taking any point that is not on the boundary and testing whether it works in the inequality. If it works, then that is the side required. If it does not work, you want the other side.

Example

Show the region $3x + 2y < 12$.

Draw the line $3x + 2y = 12$.

Test a point not on the line. The origin is always a good choice if possible, as 0 is an easy number to test. Put the value of $(0, 0)$ into the inequality, i.e. $x = 0$ and $y = 0$. Is it true that $3(0) + 2(0) < 12$?

The answer is yes, so the origin is in the side of the boundary line that we want.

Shade it in.

graphs

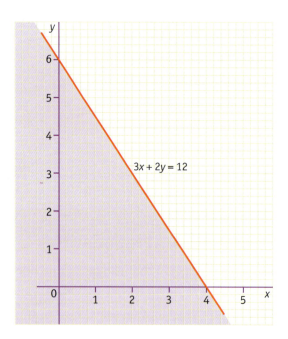

More than one inequality

When we are involved in showing a region that satisfies *more than one* inequality, it is clearer to *shade out* the regions we *do not want*, so that all that is left blank is the region that satisfies all the inequalities we are using.

Linear programming

Linear programming uses the graph techniques covered above to solve problems. This is best illustrated by an example.

Example

A man decides to buy some boats to use on his boating lake. He decides to buy at least 3 but no more than 7 canoes and at least 4 dinghies. He wants at least 12 boats. He is prepared to spend up to £600. Each canoe will cost £60 and each dinghy £40. Use graphical methods to find:

(a) the options that satisfy the conditions, and which option is the cheapest
(b) the largest number of boats that he can buy

Write down the inequalities that describe the situations above:

Number of boats:	canoes	$3 \leq c \leq 7$	(1)
	dinghies	$d \geq 4$	(2)
Money		$60c + 40d \leq 600$	
which cancels to		$3c + 2d \leq 30$	(3)
Number of boats		$c + d \geq 12$ (4)	

Now draw these inequalities on a grid. (The size of the grid is usually given in an exam.)

In the diagram the desired region is left unshaded. This unshaded region is often referred to as 'the feasible region'.

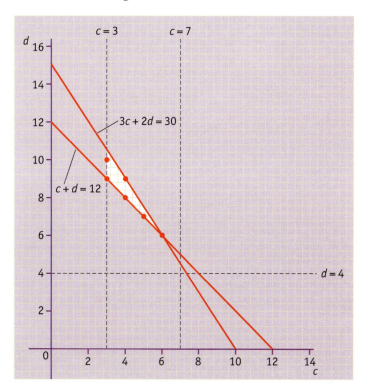

Within the region that satisfies all the inequalities there are several combinations of c and d. These are marked with dots. To answer the problem about the costs and the numbers of boats, we inspect the dots and their 'values'.

(a) The possibilities are:

Canoes	3	3	4	4	5	6
Dinghies	9	10	8	9	7	6

The cheapest option will normally be found in one of the 'corners', so these are the places to check. In this case it is £540, which is 3 canoes and 9 dinghies.

(b) The largest number of boats he could buy is 13, which could be either:

      3 canoes and 10 dinghies   total cost £580

or 4 canoes and 9 dinghies   total cost £600

Test yourself

1 Draw a pair of graphs to solve the simultaneous equations:
$$3x + 5y = 12$$
$$y = 4x - 1$$

2 Draw the following graphs:

(a) $y = 3x - 1$ $-1 \le x \le 2$

(b) $3x + 4y = 24$ $-1 \le x \le 9$

(c) $y = x^2 - 3x$ $-1 \le x \le 4$

(d) $y = \dfrac{10}{x}$ $-10 \le x \le 10$

(e) $y = \sin 2x$ $0 \le x \le 360°$

(f) $y = \sqrt{x}$ $0 \le x \le 10$

3 (a) Sketch the graph of $y = x^2$.

(b) On the same diagram, sketch graphs to show:
     (i) $y = x^2 - 3$ (ii) $y = (x - 1)^2$

4 Write down the gradient of a straight line joining each of the following pairs of points:

(a) $(2, 3)$, $(3, 6)$

(b) $(3, 6)$, $(5, 10)$

(c) $(4, 2)$, $(7, 1)$

5 (a) What does the gradient of a distance–time graph tell us?

(b) What does the gradient of a velocity–time graph tell us?

(c) What does the area under a velocity–time graph represent?

6 Draw graphs to show the region satisfying both $x + 2y < 10$ and $3x - y < 1$. Use your graphs to find the maximum value of $x + y$, where x and y are both integers.

7 (a) Draw the graph for $y = x^2 - x - 2$.

(b) Use your graph to solve the equation $x^2 - x - 2 = 1$.

(c) Use your graph to solve the equation $x^2 - x = 4$.

Practice questions

1 200 tonnes of industrial waste is accidentally spilt onto a beach. During the clean up, 40% of the waste remaining is cleared each week.

After w weeks, T tonnes of industrial waste remain on the beach.

(a) Complete the table giving the values of T correct to 1 decimal place.

Number of weeks (w)	0	1	2	3	4	5
Amount of waste left (T)	200			43.2		

(b) Draw a graph to represent the information in the table.

(c) Use your graph to help estimate when there will be less than 10 tonnes of waste left on the beach.

2 The graph below shows the distance, in metres, travelled by a car during the first minute of a race.

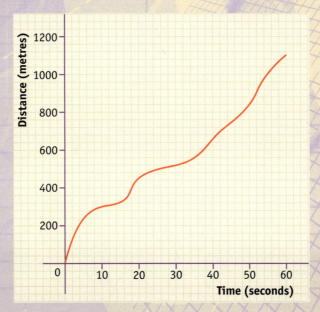

(a) Use the graph to estimate the greatest speed, in metres per second, reached by the car during this minute.

(b) Calculate the average speed, in metres per second, of the car during the minute.

This graph shows the speed, in metres per second, of the car during the last minute of the race.

(c) Use the graph to estimate the distance travelled, in metres, by the car during the last minute.

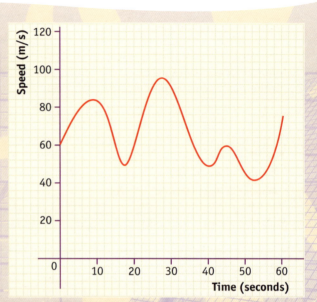

3 (a) Draw a graph of $y = 2x^2 - 5x$ for $-2 \leq x \leq 4$.

(b) Find the coordinates of the point on the graph that has a gradient of 4.

4 Use a graphical method to solve the equation $\dfrac{1}{x} = x^2 - 2$.

5 The graph shown here is $y = \dfrac{1}{x}$.

(a) Draw a sketch to illustrate
$$y = \frac{1}{x - 2}$$

(b) Draw a sketch to illustrate
$$y = \frac{1}{x} + 3$$

(c) Draw a sketch to illustrate
$$y = \frac{1}{x - 2} + 3$$

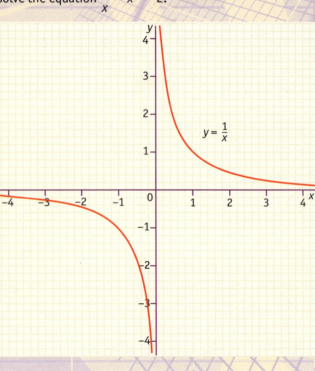

48

6 (a) Draw the graph of $y = 2 + \sin 3x$ for $0° \leq x \leq 180°$.

(b) Use your graph to solve the equation $\sin 3x = 0.4$. Show clearly on your graph how you found the solution.

7 Certain bacteria grow very quickly. One type will double its number every 3 hours.

Imagine one cell of this bacteria is placed in a host. Complete the table below, which shows the total number of cells of the bacteria in the host at various times.

Time (hours)	0	3	6	9	12	15	18
Number of cells	1						

(a) Draw a graph to represent the information above.

(b) Write an equation connecting the number of cells, C, with the time, T, after placing one cell in the host.

These particular bacteria only become dangerous to the host if there are over 1000 cells present.

(c) After how many hours will the bacteria be dangerous to the host?

8 The water resistance to a boat at various speeds was measured and recorded.

Speed (knots), S	5	10	15	20	25
Resistance (units), R	5	7.1	8.7	10	11.2

(a) Which of these graphs best fits the information given?

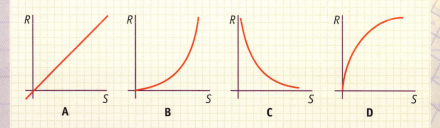

A B C D

(b) Which of these equations describes the graph you have chosen?

$R = kS^2$ $R = kS$ $R = k\sqrt{S}$ $R = k/S$

(c) Find the value of k for the equation you have chosen.

9 (a) What is the equation of the line that is parallel to the line $y = 3x - 4$ and passes through the point (0, 5)?

(b) What is the equation of the line that is parallel to the line $y = 3x - 4$ and passes through the point (2, 5)?

10 A factory that makes 'Widgits' is to be re-equipped. There are two types of machine used and the replacement specifications are:

Type	Cost	Daily output	No. of people per machine
Cutter	£500	30 widgits	2
Grinder	£1500	40 widgits	1

The company does not wish to spend more than £40 000 on the new equipment and the factory can accommodate no more than 30 machines and no more than 50 workers.

(a) How many of each machine should be installed to give the greatest daily output?

(b) What is the cost of your solution?

Shape

The area of a trapezium

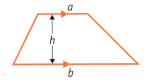

The area of a trapezium is found by multiplying the average length of the parallel sides by the vertical difference between them.

i.e. area $= h\dfrac{(a + b)}{2}$

For example:

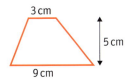

The area of the trapezium shown is:

$$5 \times \frac{(3 + 9)}{2}$$

which equals $30\,\text{cm}^2$.

Measuring sectors

A sector is part of a circle.

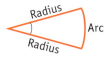

Each sector has two radii that subtend an angle (the angle of the sector).

· The part of the circumference shown here is called the arc.

If a circle is divided into two sectors, then the large one is called the major sector and the small one is called the minor sector.

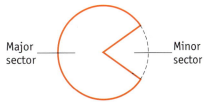

Likewise the arcs in these sectors are called the major arc and the minor arc.

These sectors are simple fractions of the whole circle. You can find the fraction of the whole circle using the angle of each sector.

For example, if the radii r of a sector subtend an angle θ then

$$\text{arc length} = \frac{\theta}{360°} \times 2\pi r \quad (2\pi r \text{ is the circumference of the circle})$$

$$\text{sector area} = \frac{\theta}{360°} \times \pi r^2 \quad (\pi r^2 \text{ is the area of the circle})$$

Example

Find the arc length and the area of this sector:

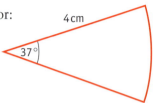

The sector angle is 37° and the radius is 4 cm.

$$\text{Arc length} = \frac{37°}{360°} \times \pi \times 2 \times 4 = 2.6 \text{ cm (1 d.p.)}$$

$$\text{Sector area} = \frac{37°}{360°} \times \pi \times 4^2 = 5.2 \text{ cm}^2 \text{ (1 d.p.)}$$

The volume of a prism

A prism is a solid shape that has the same cross-section running all the way through it.

The volume of a prism is found by multiplying the area of the regular cross-section by its length (or height if it is standing on its end).

i.e. volume of prism = area of cross-section × length

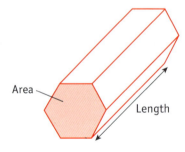

A cylinder is a prism with area of cross-section = πr^2. Hence its volume is given by $\pi r^2 h$, where h is its height.

Example

The diagram shows a prism with a triangular cross-section.

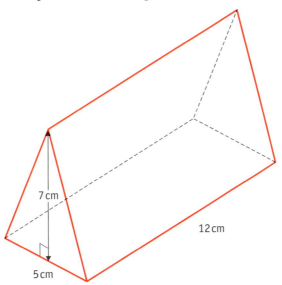

7 cm

12 cm

5 cm

The area of the cross-section is the area of the end triangle, which is:

$$\frac{\text{height} \times \text{base}}{2} = \frac{5 \times 7}{2} = 17.5 \, \text{cm}^2$$

The volume is the area of cross-section × length, which is:

$$17.5 \, \text{cm}^2 \times 12 \, \text{cm} = 210 \, \text{cm}^3$$

The volume of a pyramid

A pyramid is a three-dimensional shape with a base. All the sides rise up from the base to form a point at the top. (The top is called the 'apex'.)

The base can be any shape, but it is usually a triangle, a rectangle or a square. (It can be a circle — you meet that in the next section.)

$$V = \frac{1}{3} Ah$$

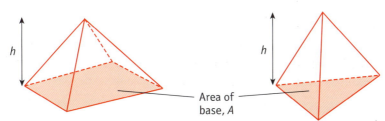

h

h

Area of base, A

Example

A pyramid has a rectangular base of sides 4 cm and 5 cm. Its vertical height is 6 cm. Calculate the volume of the pyramid.

The area of the base is $4 \times 5 = 20\,\text{cm}^2$

Volume $= \frac{1}{3} \times 20 \times 6 = 40\,\text{cm}^3$

Cones and spheres

The cone

A cone is a pyramid with a circular base. The formula for the volume of a cone is $\frac{1}{3}$ of the base area multiplied by the vertical height:

$V = \frac{1}{3}\pi r^2 h$

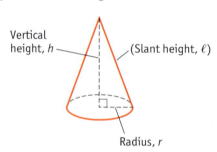

Vertical height, h — (Slant height, ℓ)

Radius, r

The formula for the curved surface area of the cone is:

curved surface area $= \pi r \ell$

where ℓ is the slant height of the cone.

Note: There is a link between the radius, the vertical height and the slant height of a cone. As you can see in the diagram, they form a right-angled triangle. From Pythagoras's theorem, which you will find in the next section:

$\text{radius}^2 + (\text{vertical height})^2 = (\text{slant height})^2$

Example

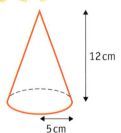

12 cm

5 cm

For the cone in the diagram, calculate:
(i) the volume
(ii) the total surface area

(i) Volume $= \frac{1}{3}\pi r^2 h = \frac{1}{3} \times \pi \times 25 \times 12 = 314\,\text{cm}^3$ (3 s.f.)

(ii) To find the curved surface area, we need to calculate the slant height:

$$\ell^2 = 5^2 + 12^2 = 169$$

$$\ell = \sqrt{169} = 13\,\text{cm}$$

Total surface area = curved surface area + base area

$$= \pi r \ell + \pi r^2$$

$$= \pi \times 5 \times 13 + \pi \times 25$$

$$= 204.20352 + 78.539816$$

$$= 283\,\text{cm}^2 \text{ (3 s.f.)}$$

The sphere

A sphere is a round ball. Examples of spheres are footballs, marbles, tennis balls, snooker balls, ball bearings and so on.

The formula for finding the volume of a sphere with radius *r* is:

$$V = \tfrac{4}{3}\pi r^3$$

The formula for finding the surface area of a sphere with radius *r* is:

$$A = 4\pi r^2$$

Example

For a sphere of radius 5 cm, calculate:
(i) the volume
(ii) the surface area

(i) Volume $= \tfrac{4}{3} \times \pi \times 5^3 = 524\,\text{cm}^3$ (3 s.f.)

(ii) Surface area $= 4 \times \pi \times 5^2 = 314\,\text{cm}^2$ (3 s.f.)

Density

Density is the amount of weight per unit volume, usually given in grams per cm³.

You need to remember that weight is commonly used in mathematics examination questions whereas in science it is always referred to as mass.

$$\text{density} = \frac{\text{weight}}{\text{volume}}$$

This connection can be remembered with a triangle:

This reminds you that:

weight is   volume × density
density is   weight ÷ volume
volume is   weight ÷ density

Example 1

A piece of metal weighing 25 g has a volume of 8 cm³. What is its density?

$$\text{Density} = \frac{25}{8} = 3.1 \, \text{g/cm}^3$$

Example 2

Find the weight of a rock that has a volume of 29 cm³ and a density of 1.7 g/cm³.

$$\text{Weight} = 29 \times 1.7 = 49.3 \, \text{g}$$

Similarity

Triangles are similar if their corresponding angles are equal. Their corresponding sides are then in the same ratio.

Example

Find the sides marked x and y in these triangles (not drawn to scale).

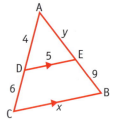

Triangles AED and ABC are similar, so the corresponding sides CB and DE, and AC and AD, are in the same ratio, that is

$$\frac{CB}{DE} = \frac{AC}{AD}$$

Thus

$$\frac{x}{5} = \frac{10}{4}$$

$$\Rightarrow x = \frac{10 \times 5}{4} = 12.5$$

Using the corresponding sides AB and EB, and AC and DC gives:

$$\frac{AB}{EB} = \frac{AC}{DC}$$

Thus

$$\frac{y + 9}{9} = \frac{10}{6}$$

$$y + 9 = \frac{9 \times 10}{6}$$

$$y = 15 - 9 = 6$$

Example

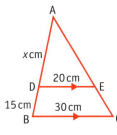

Find the value of x in this triangle.

We know that triangle ABC is similar to triangle ADE.

Splitting the triangles up may help us to see what will be awkward (and often missed).

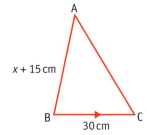

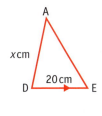

So the equation will be $\dfrac{x + 15}{x} = \dfrac{30}{20}$

Multiplying through by x and then by 20 gives

$$20x + 300 = 30x$$
$$300 = 10x \Rightarrow x = 30\,\text{cm}$$

As long as you recognise this situation and are comfortable with the algebra, it is not as difficult as it may first appear. But do draw two triangles if this helps you.

Ratios of areas and volumes of similar figures

You should learn these ratios for similar figures:

length ratio $x{:}y$
area ratio $x^2{:}y^2$
volume ratio $x^3{:}y^3$

This information can be useful in solving problems, as shown in the following examples.

Example 1

A model yacht is made to $\frac{1}{25}$ of the size of the real yacht. The area of the sail of the model is $140\,\text{cm}^2$. What is the area of the sail of the real yacht?

length ratio = 1:25 $(x{:}y)$

area ratio = 1:625 $(x^2{:}y^2)$

area of real sail = $625 \times$ area of model sail
$$= 625 \times 140$$
$$= 87\,500\,\text{cm}^2 = 8.75\,\text{m}^2$$

Example 2

A bottle has a base radius of $5\,\text{cm}$ and holds $750\,\text{cm}^3$. A similar bottle has a base radius of $4\,\text{cm}$. What is the volume of the smaller bottle?

The length ratio is 5:4

The volume ratio is $5^3{:}4^3 = 125{:}64$

The ratio of $\dfrac{\text{volume of small bottle}}{\text{volume of large bottle}} = \dfrac{64}{125} = \dfrac{v}{750}$

$$\Rightarrow v = \frac{64 \times 750}{125} = 384\,\text{cm}^3$$

Example 3

The cost of a tin of paint of height $15\,\text{cm}$ is £2.30 and is proportional to the volume of the tin. Its label has an area of $34\,\text{cm}^2$.
(a) How much will a similar tin of height $20\,\text{cm}$ cost?
(b) What will be the area of the label on the bigger tin?

(a) The cost of the paint is proportional to the volume.

length ratio = 15:20 = 3:4

volume ratio = $3^3{:}4^3 = 27{:}64$

$$\frac{\text{price of large tin}}{\text{price of small tin}} = \frac{64}{27} = \frac{P}{2.30}$$

$$\Rightarrow P = \frac{64 \times 2.3}{27} = £5.45$$

(b) Area ratio = $3^2{:}4^2 = 9{:}16$

$$\frac{\text{large label area}}{\text{small label area}} = \frac{16}{9} = \frac{A}{34}$$

$$\Rightarrow A = \frac{16 \times 34}{9} = 60.4\,\text{cm}^2 \text{ (3 s.f.)}$$

Test yourself

1 Calculate the volumes of the following prisms.

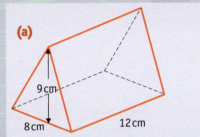 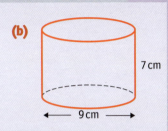

2 (a) Calculate the weight of a solid cuboid with dimensions 10 cm by 6 cm by 3.5 cm and a density of 1.8 g/cm³.

(b) A stone has a volume of 13 cm³ and weighs 31.5 g. What is its density?

3 (a) Find the arc length and area of a sector of radius 6 cm and angle 18°.

(b) Find the area of a sector that has radius 5 cm and arc length 3.5 cm.

4 Calculate the volume and the curved surface area of each of these cones:

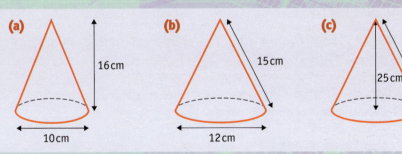

5 Calculate the volume of each of these pyramids:

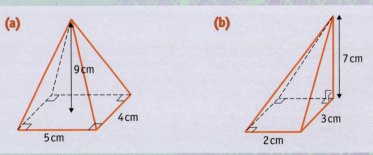

6 (a) Calculate the volume of a sphere with surface area $100\,cm^2$.

(b) Calculate the surface area of a sphere with a volume of $250\,cm^3$.

7 Calculate the values of x and y in the figures below.

(a)

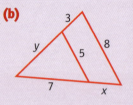

(b)

Practice questions

1 A square-based pyramid has its vertical height of 18 cm above the centre of the square base which has an area of $160\,cm^2$.

Calculate the volume of the pyramid.

2 Find the volume of a sphere that has the same surface area as the curved surface area of a cone with radius 5 cm and a height of 14 cm.

3 What is the weight of a hollow metal sphere with inner diameter 20 cm and thickness 15 mm, given that the density of the metal is $4.3\,g/cm^3$?

4 The diagram shows a swimming pool.

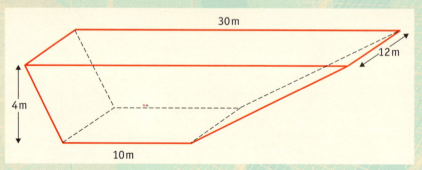

What is the volume of the pool?

5 A cylindrical can of orange juice has a base radius of 4 cm and a volume of 500 cm³. Calculate the height of the can. Give your answer correct to 3 significant figures.

6 Jade has a pan with a circular base of radius 10 cm and depth 13 cm. It is filled with water to a depth of 11 cm. She places a solid metal cylinder of radius 4.8 cm and height 12 cm on its side into the water. Does the water level rise above the pan?

7

Find the lengths marked *x* and *y*.

8 (a) Breakfast cereal comes in similar packets in regular size and small size as shown.

  (i) Calculate the height of the small packet.
  (ii) Calculate the depth of the regular packet.

(b) Two jars of coffee are similar in shape. The small jar has a net weight of 85 grams, the large one has a net weight of 250 grams.

The height of the small jar is 9 cm. Calculate the height of the large jar.

Note: For questions 9 and 10, you need to know Pythagoras's theorem (p. 63).

9

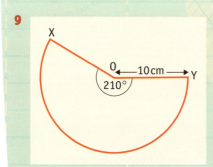

XY is the arc of a sector of radius 10 cm with an angle of 210°.

OX is joined to OY to form a cone. Calculate the volume inside this cone.

10 A lamp shade is made as part of a cone, as shown.

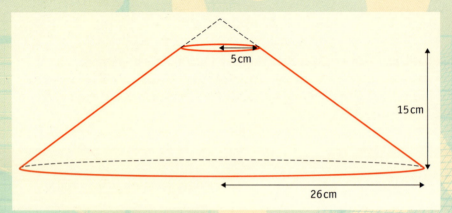

Calculate how much material is needed to cover the shade, that is, the curved surface area of the shade.

Pythagoras and trigonometry

Pythagoras's theorem

In any right-angled triangle, the sum of the squares of the lengths of the two short sides is equal to the square of the hypotenuse.

This can be written more conventionally as:

$a^2 + b^2 = c^2$

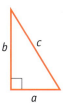

Example 1

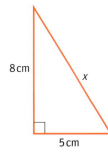

Find the length of the hypotenuse.

$$x^2 = 8^2 + 5^2 = 64 + 25 = 89$$
$$x = \sqrt{89} = 9.4 \text{ cm (1 d.p.)}$$

Example 2

Find the length of the small side x.

$$x^2 + 7^2 = 12^2$$
$$x^2 = 12^2 - 7^2 = 144 - 49 = 95$$
$$x = \sqrt{95} = 9.7 \text{ cm (1 d.p.)}$$

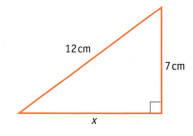

Right-angled trigonometry

Right-angled trigonometry is based on a right-angled triangle like this:

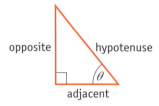

The relationships are all built around these sides as defined:

$$\text{sine}\,\theta = \frac{\text{opposite}}{\text{hypotenuse}} \qquad \text{cosine}\,\theta = \frac{\text{adjacent}}{\text{hypotenuse}} \qquad \text{tangent}\,\theta = \frac{\text{opposite}}{\text{adjacent}}$$

This is best learnt by some mnemonic, such as:

Silly **O**ld **H**itler **C**ouldn't **A**dvance **H**is **T**roops **O**ver **A**frica

which, taking the first letter from each word, reminds us that

$$S = \frac{O}{H} \qquad C = \frac{A}{H} \qquad T = \frac{O}{A}$$

Shorthand

The full trigonometric names are shortened to:

sine = sin   cosine = cos   tangent = tan

and you will find buttons for these on your calculator.

> **Important:**
>
> Make sure that your calculator is working in degrees. You can usually tell this by a small 'D' or 'DEG' in the display.
>
> You need to put your calculator into 'degree mode' before you start working on sines and cosines.
>
> Depending on the type of calculator you have, this may be done by pressing the keys MODE 4 or by pressing the key DRG until D or deg is in the display.

Solving problems

If you have a problem to solve that involves trigonometry:

- first, draw a triangle
- add the given information
- decide which bit of trigonometry you need

Example 1

Find the angle θ in the triangle below.

12 cm   5 cm

θ

Note that the opposite side is 5 cm and the hypotenuse is 12 cm.

Opposite and hypotenuse leads us to the sine of the angle:

$$\sin \theta = \frac{\text{opposite}}{\text{hypotenuse}} = \frac{5}{12} = 0.4166666$$

To find the angle, you do not actually need to work out the 0.4166 etc. You can find the 'inverse sin ($\sin^{-1}$)' on the calculator by:

$5 \div 12 =$ [shift] [sin]

Check that you are able to get the answer of 24.6° (rounded).

Warning: there are many different types of calculator around, so get used to the one you will use in the examination. (Your calculator may use [2nd F] or something else instead of shift, but it is usually the top left-hand button.)

Trouble shooting

If you are unable to get the correct answer to the above problem, then check these things:

Wrong answer	Cause	Remedy
0.4297754	You are in RAD mode	Put the calculator into DEG mode, either by keying in [MODE] [4] or by pressing [DRG] until you get D or deg in the display
27.360354	You are in GRAD mode	As above
$- E -$	You probably divided 12 by 5 instead of the correct way round	Divide 5 by 12 first, press = and then access $\sin^{-1}$
72.6 or 25.1	You have used cos or tan instead of using sin	Use the sin key

Example 2

Find the side marked x in the triangle.

Look to see which kind of side is wanted — it is opposite to the known angle. So we are looking for the opposite, with the hypotenuse and an angle known.

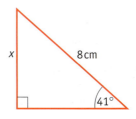

x 8 cm

41°

Use

$$\sin \theta = \frac{\text{opposite}}{\text{hypotenuse}}$$

so

$$\sin 41° = \frac{\text{opposite}}{8}$$

opposite $= 8 \times \sin 41° = 5.2\,\text{cm}$ (rounded)

Example 3

Find the hypotenuse of the triangle.

Note that we are given the side opposite to the known angle.

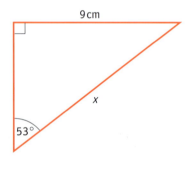

Use

$$\sin \theta = \frac{\text{opposite}}{\text{hypotenuse}}$$

so

$$\sin 53° = \frac{9}{\text{hypotenuse}}$$

$$\text{hypotenuse} \times \sin 53° = 9$$

$$\text{hypotenuse} = \frac{9}{\sin 53°} = 11.3\,\text{cm}\ (1\ \text{d.p.})$$

Example 4

Find the angle θ in the triangle.

We see that the adjacent side is 5 cm and the hypotenuse is 8 cm.

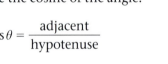

We use the cosine of the angle:

$$\cos \theta = \frac{\text{adjacent}}{\text{hypotenuse}}$$

$$\cos \theta = \frac{5}{8} = 0.625$$

Use your calculator and $\cos^{-1}$ to find $\theta = 51.3°$ (rounded)

Example 5

Find the side marked x in the triangle.

Note that we are looking for the opposite, but the adjacent side and an angle are known. So we use tangent:

$$\text{tangent} = \frac{\text{opposite}}{\text{adjacent}}$$

so

$$\tan 73° = \frac{\text{opposite}}{8}$$

$$\text{opposite} = 8 \times \tan 73°$$
$$= 26.2 \, \text{cm (rounded)}$$

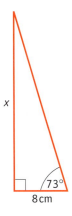

Pythagoras and trigonometry in three dimensions

Remember that most trigonometry and Pythagoras problems at GCSE do not come in the form of a nice straightforward triangle. Usually they are part of a problem set in a real-life context. In this case, *you must draw a triangle*. Sometimes the triangle is part of the picture that accompanies the problem. Even in this case, *redraw the triangle*. This will avoid any confusion and help you to identify the correct sides and then get the required ratio.

Pythagoras's theorem in three dimensions

At the higher tier, problems use Pythagoras's theorem in three dimensions. These are usually accompanied by a diagram that serves as a guide to lengths and angles.

Example

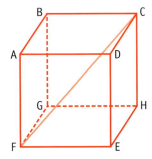

Find FC from

$$(\text{FC})^2 = (\text{CH})^2 + (\text{FH})^2$$

but find FH from

$$(\text{FH})^2 = (\text{FE})^2 + (\text{EH})^2$$

The following rules may help you.

- Find a right-angled triangle that connects two known lengths and the length you want to find.

- Redraw this triangle as a right-angled triangle, marking on the lengths that you know and the length that you want to find.
- Decide which type of Pythagoras problem it is. (Remember there are only two basic types — find one of the short sides or find the hypotenuse.)
- Solve the problem.

The sine rule

For any triangle labelled as:

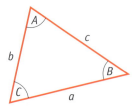

it is true that:

$$\frac{a}{\sin A} = \frac{b}{\sin B} = \frac{c}{\sin C}$$

The rule can be rearranged to put all the sines at the top:

$$\frac{\sin A}{a} = \frac{\sin B}{b} = \frac{\sin C}{c}$$

Note that in many situations, triangles are not labelled. The way to combine sides and angles is to remember that the side is divided by the sine of the opposite angle.

If you are calculating a side, use the sine rule with 'sides on top'.

If you are calculating an angle, use the sine rule with 'sines on top'.

Example 1

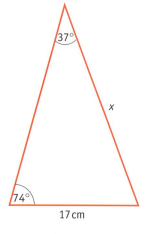

Find the value of x.

Using the sine rule (sides on top):

$$\frac{x}{\sin 74°} = \frac{17}{\sin 37°}$$

$$\Rightarrow \quad x = \frac{17 \times \sin 74°}{\sin 37°}$$

$$= 27.2 \, \text{cm (3 s.f.)}$$

Example 2

Find the value of x.

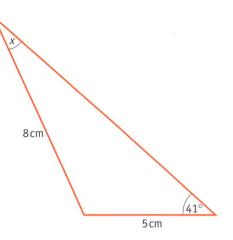

Using the sine rule (sines on top):

$$\frac{\sin x}{5} = \frac{\sin 41°}{8}$$

$$\Rightarrow \quad \sin x = \frac{5 \times \sin 41°}{8} = 0.4100$$

$$\Rightarrow \quad x = 24.2° \text{ (3 s.f.)}$$

Area using sines

For any triangle ABC with sides labelled a, b and c as usual (see page 68), the area of the triangle is found by:

$$\text{area} = \frac{1}{2}ab\sin C$$

Example

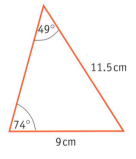

Find the area of the triangle.

The angle between the two known sides is $180 - (74 + 49) = 57°$.

$$\text{Area} = \frac{1}{2} \times 11.5 \times 9 \times \sin 57° = 43.4\,\text{cm}^2 \text{ (rounded)}$$

The cosine rule

For this triangle, the cosine rule states that the following are always true:

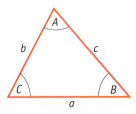

$$a^2 = b^2 + c^2 - 2bc\cos A$$

$$b^2 = a^2 + c^2 - 2ac\cos B$$

$$c^2 = a^2 + b^2 - 2ab\cos C$$

Look at the symmetry of the rule and learn how it works, using two sides and the angle between them.

The formula can be rearranged to give the cosine rule to find angles:

$$\cos A = \frac{b^2 + c^2 - a^2}{2bc}$$

or

$$\cos B = \frac{a^2 + c^2 - b^2}{2ac}$$

and

$$\cos C = \frac{a^2 + b^2 - c^2}{2ab}$$

Remember:
- find a side using the cosine rule, where $a^2 = \ldots$
- find an angle using the cosine rule, where $\cos A = \ldots$

Example 1

Find x.

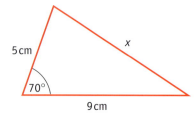

By the cosine rule:

$$x^2 = 5^2 + 9^2 - 2 \times 5 \times 9 \times \cos 70°$$

$$x^2 = 75.218$$

$$x = 8.67 \text{ cm (3 s.f.)}$$

Example 2

Find θ.

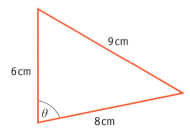

By the cosine rule:

$$\cos \theta = \frac{6^2 + 8^2 - 9^2}{2 \times 6 \times 8} = 0.1979$$

$$\theta = 78.6° \text{ (3 s.f.)}$$

Test yourself

1 For each triangle below, calculate the length marked *x*.

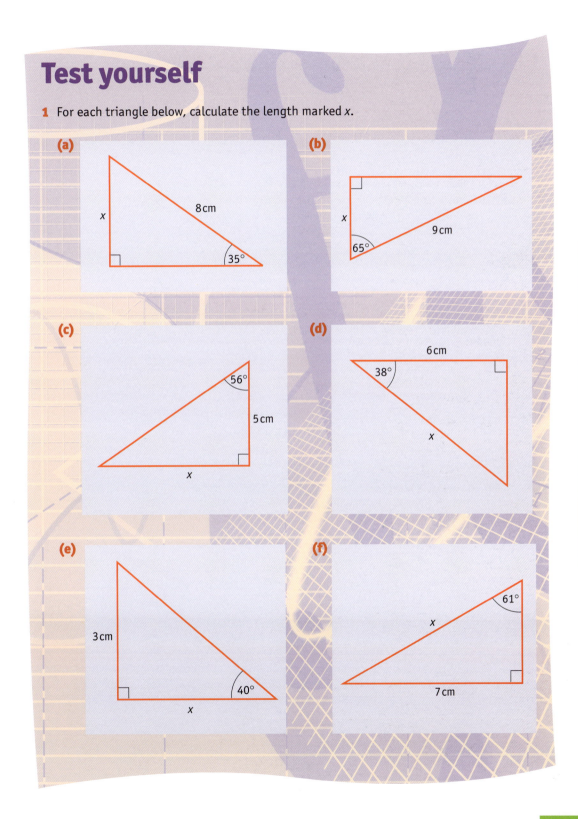

(a)

x

8 cm

35°

(b)

x

9 cm

65°

(c)

56°

5 cm

x

(d)

6 cm

38°

x

(e)

3 cm

40°

x

(f)

61°

x

7 cm

2 For each triangle below, calculate the angle marked θ.

(a)

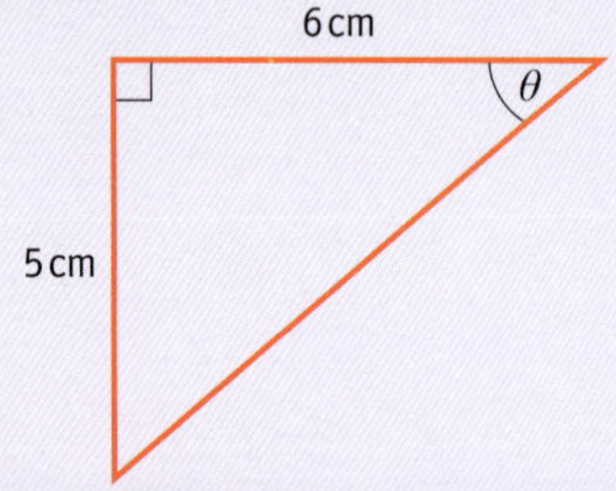

6 cm

5 cm

θ

(b)

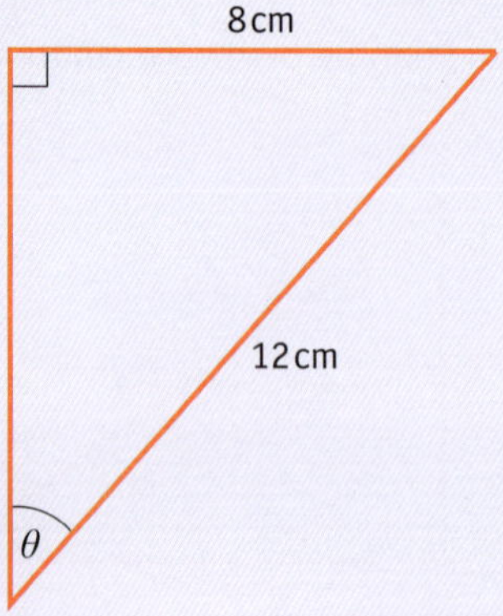

8 cm

12 cm

θ

(c)

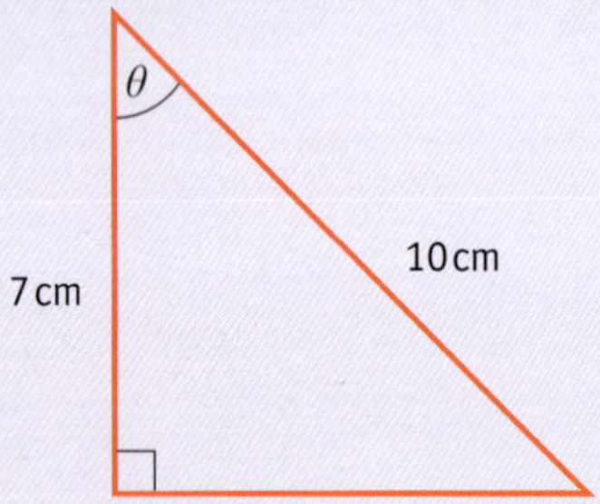

θ

7 cm

10 cm

3 For each triangle below, calculate the length marked x.

(a)

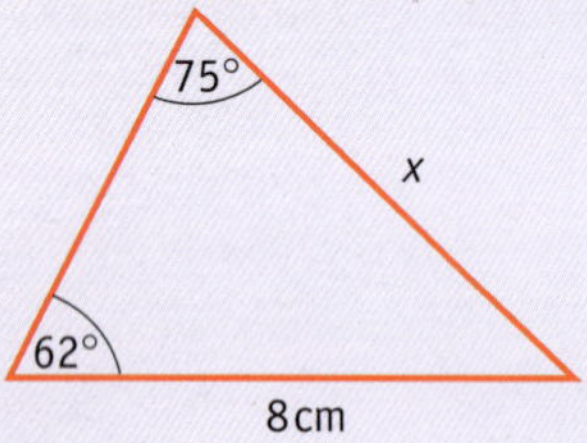

75°

x

62°

8 cm

(b)

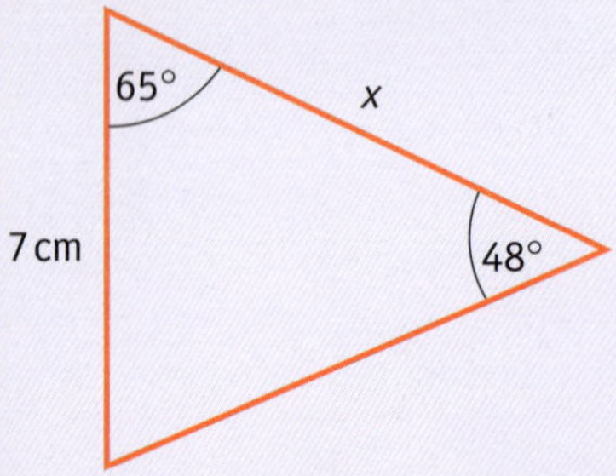

65°

x

7 cm

48°

(c)

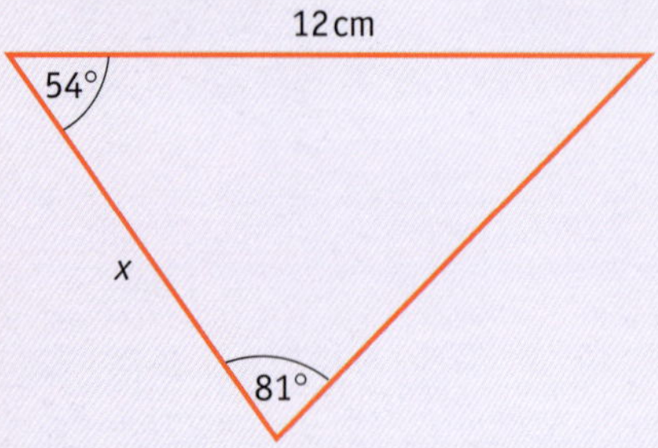

12 cm

54°

x

81°

4 For each triangle below, calculate the value of *x*.

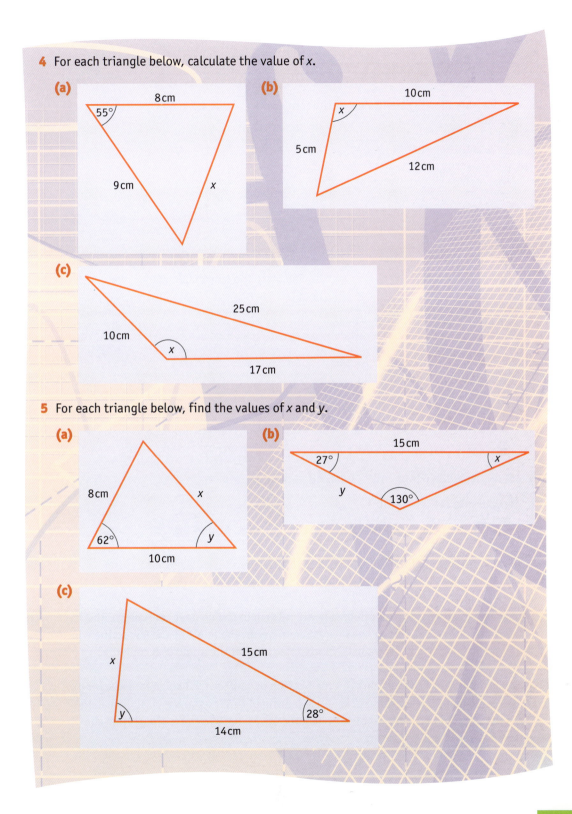

(a)

8 cm
55°
9 cm
x

(b)

10 cm
x
5 cm
12 cm

(c)

25 cm
10 cm
x
17 cm

5 For each triangle below, find the values of *x* and *y*.

(a)

8 cm
x
62°
y
10 cm

(b)

15 cm
27°
x
y
130°

(c)

15 cm
x
y
28°
14 cm

Practice questions

1 The diagram shows the end view of the framework for a building.

Calculate the length AB.

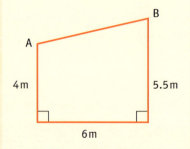

2 Calculate the length CD.

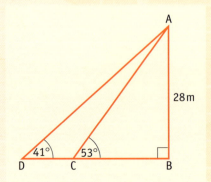

3 ABCD is a parallelogram. AB = 9 cm, AD = 5 cm, AC = 6 cm.

  (a) Calculate the length BD.

  (b) Calculate the obtuse angle between the diagonals AC and BD.

4 What is the length of the longest stick that can be put inside a box measuring 4 cm by 13 cm by 27 cm?

5 In the triangle ABC, the length of side AB is 34 cm to the nearest centimetre. The length of side AC is 27 cm to the nearest centimetre. The angle at C is 53° to the nearest degree. What is the smallest possible size that the angle at B could be?

6 A helicopter flies from heliport H, on a bearing of 134°, for 5.7 km before reaching an injured man. It then flies 8.3 km on a bearing of 078° to the hospital. Calculate the distance and bearing that the helicopter will fly back to the heliport from the hospital.

7 Find the area of the parallelogram ABCD where AB = 4 cm, BC = 3 cm and angle ABC = 149°.

8 Sam is trying to find the height of a tall building. He measures the angle of elevation of the top of the building to be 37°. Then he walks 30 metres further away from the building along level ground, and the angle of elevation is now 27°. How high is the building?

9 The total perimeter of a sector subtending an angle of 57° is 20 cm.

   (a) What is the radius of the sector?

   (b) A shield is made to the shape and size of this sector, and is 3 mm thick. What is the volume of the shield?

10 Triangle ABC has an area of 8.5 cm². Length AB is 3.1 cm and length AC is 8.2 cm.

   (a) Calculate the angle at A.

   (b) Calculate the size of the angle at B.

Geometry and construction

Parallel lines

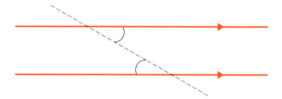

The diagram shows two parallel lines (the arrows indicate that they are parallel).

The line cutting through the parallel lines is called a transversal. You can see in the diagram that this creates equal angles.

 The correct name for these equal angles is **alternate angles**, but they are often called Z angles because the transversal and parallel lines make a Z shape.

A similar situation is shown in the following diagram:

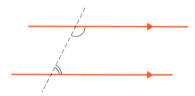

 Here the two angles shown add up to 180°. Two angles like this are called **allied angles**. They are also called F angles, because the lines make an F shape.

Angles in a polygon

- For an *N*-sided polygon the interior angles add up to $180° \times (N - 2)$.
- Regular polygons have every interior angle equal and each side the same length.

- The exterior angle and the interior angle add up to 180°.

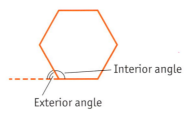

Interior angle

Exterior angle

- The exterior angle of a regular N-sided polygon = 360° ÷ N.
- The interior angle of a regular N-sided polygon
  = 180° − exterior angle
  = 180° − (360° ÷ N)

Example 1

What is the sum of the interior angles of a regular nine-sided polygon?

angle sum = 180° × (9 − 2) = 1260°

Example 2

What are the exterior and the interior angles of a regular polygon with 12 sides?

exterior angle = 360° ÷ 12 = 30°

interior angle = 180° − 30° = 150°

Angles in a circle

You need to be familiar with these three circle theorems. Try proving them for yourself.

- An angle at the centre of a circle is twice any angle at the circumference subtended by the same arc.

$\angle AOB = 2 \times \angle ACB$

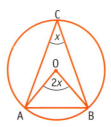

- Every angle in a semicircle subtended by the diameter is a right angle.

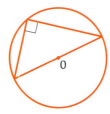

- From a chord AB, any angle subtended at the
  circumference, in the same segment of the circle,
  will have the same value.

  $\angle AC_1B = \angle AC_2B = \angle AC_3B$

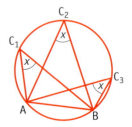

Cyclic quadrilaterals

A quadrilateral with its four vertices lying on the
circumference of a circle is called a cyclic
quadrilateral. In a cyclic quadrilateral, the
opposite angles add up to 180°.

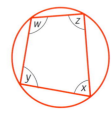

$w + x = 180°$
$y + z = 180°$

Angles on a tangent

A tangent is a straight line drawn so that it touches a circle at only one
point (however long you make the line). This point is called the point of
contact.

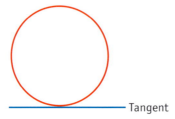

Tangent

There are two important facts about tangents to circles:

- The tangent to a circle is perpendicular to the radius drawn to the point of
  contact.
- Tangents to a circle drawn from an external point to their points of contact are
  equal in length. The line joining the external point to the centre of the circle
  bisects the angle between the tangents.

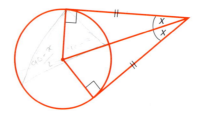

Alternate segment theorem

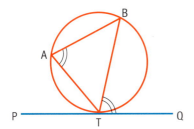

PTQ is the tangent to the circle at T. The segment containing TAB is known as the alternate segment of ∠QTB, because it is on the other side of the chord BT from ∠QTB.

The alternate segment theorem states:

The angle between a tangent and a chord through the point of contact is equal to the angle in the alternate segment.

Example

Find the following angles:

(i) ACB
(ii) DCA
(iii) CBA

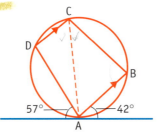

(i) ACB = 42° (alternate segments)
(ii) DCA = 57° (alternate segments)
(iii) CBA = 180° − (42° + 57°) = 81° (angles in a circle and trapezium)

Special quadrilaterals

- A **trapezium** has one pair of opposite sides parallel. The angles at the same end of the parallel lines add up to 180°.

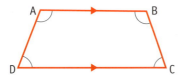

∠A + ∠D = 180°
∠B + ∠C = 180°

- A **parallelogram** has both opposite sides parallel. The opposite sides are equal in length. The diagonals of a parallelogram bisect each other. The opposite angles of a parallelogram are equal to each other, i.e. ∠A = ∠C and ∠B = ∠D.

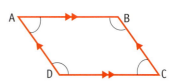

- A **rhombus** is a parallelogram with all its sides the same length. The diagonals of a rhombus bisect each other at right angles. The diagonals of a rhombus bisect the angles.

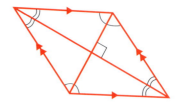

- A **kite** is a quadrilateral with two pairs of equal adjacent sides. The diagonals bisect at right angles. The angles between the long side and the short side are equal, ∠B = ∠D.

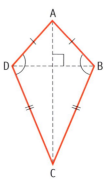

Congruence

Shapes that are identical to each other in terms of size and angles are said to be congruent to each other.

Minimum information for congruent triangles

For triangles, a certain amount of information is sufficient to indicate that they are congruent. The following examples all show congruence.

- All sides are the same, known as SSS (side, side, side):

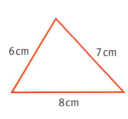
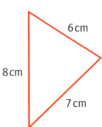

- Two sides and the angle between them are equal, known as SAS (side, angle, side):

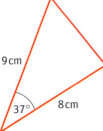
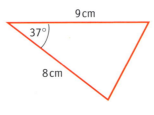

- Two angles and a corresponding side are equal, known as ASA (angle, side, angle):

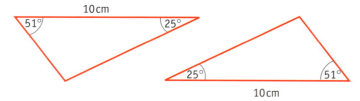

- Both triangles have a right angle with equal hypotenuse and other side, known as RHS (right angle, hypotenuse, side):

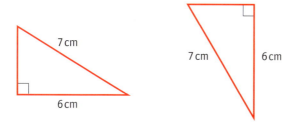

Loci

A locus (plural loci) is the movement of a point according to a rule.

Example 1

A point that moves so that it is at a distance of 5 cm from a fixed point, A, will have a locus that is a circle of radius 5 cm.

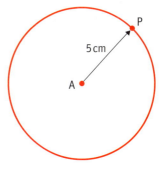

Expressing this mathematically, we would say:

the locus of the point P is such that
AP = 5 cm

Example 2

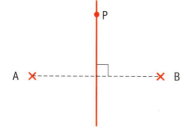

A point moves so that it is always equidistant from two fixed points A and B. (Equidistant is a word to be familiar with — it means the same distance.) This will be the perpendicular bisector of the line AB.

Expressing this mathematically, we would say:

the locus of the point P is such that AP = BP

geometry and construction

Example 3

A point that moves so that it is always 5 cm from a line AB will have a 'sausage' or 'racetrack' shape around the line.

This is difficult to express mathematically.

Practical problems

Most loci problems at GCSE are in a practical context.

Example 4

A horse is tied to a stake, on a rope that is 8 metres long, in a grassy field. What is the shape of the area that the horse can graze?

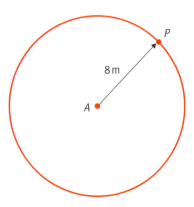

In reality, the horse may not be able to reach 8 metres if the rope is tied around its neck, but we tend not to worry about fine details like that. We 'model' the situation by saying that the horse can move around in an 8-metre circle and graze all the grass within that circle.

In this example, the locus is the whole of the area inside the circle. We can express this mathematically as:

the locus of the point P is such that AP $\leq$ 8 m

Example 5

A radio company wants to find a site for a transmitter. It wants the transmitter to be equidistant from Wuth and Brassthorpe but within 25 km of Swonton. Translated into mathematical language, this means the perpendicular bisector between Wuth and Brassthorpe and the area within a circle of radius 25 km from Swonton.

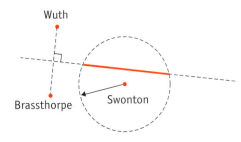

The map is drawn to a scale of 1 cm = 20 km.

So the transmitter can be sited anywhere along the thick red line.

Symmetry of three-dimensional shapes

Planes of symmetry

These are similar to lines of symmetry for two-dimensional shapes, but in two dimensions.

A solid shape may have a 'plane of symmetry', which slices the solid shape into two equal halves. One half is a reflection of the other half.

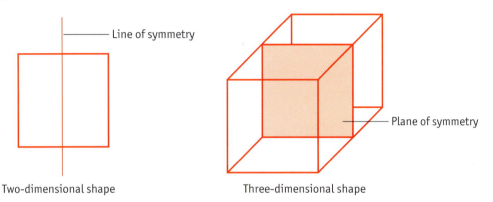

Two-dimensional shape   Three-dimensional shape

Rotational symmetry

This is similar to the rotational symmetry of flat shapes, but again in two dimensions. A solid shape may have an 'axis of symmetry'.

A solid shape has an axis of symmetry if the shape can rotate about that axis and look as if it is still in the original place.

For example, a cuboid has three axes of symmetry:

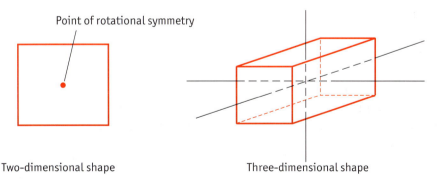

Two-dimensional shape   Three-dimensional shape

See how the shape can rotate about the axes and appear to be unaltered. For the cuboid shown, each axis of symmetry is of order two, since the solid looks unchanged at two different positions about each axis.

This is extremely difficult to show on a diagram. If you can, get a cuboid and see where these axes of rotational symmetry actually are.

Vectors

A vector is represented by a straight line of certain length and direction. We can put vectors together to form triangles and other plane figures.

The example below shows how vectors are used in geometry.

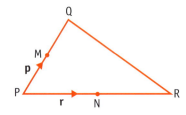

Example

PQ = **p** and PR = **r**. M and N are mid-points of PQ and PR respectively. Show that MN is parallel to QR.

We need to find vector expressions for MN and QR.

QR is the easiest:

$$QR = QP + PR = -\mathbf{p} + \mathbf{r} = \mathbf{r} - \mathbf{p}$$

$$MN = MP + PN = -\tfrac{1}{2}\mathbf{p} + \tfrac{1}{2}\mathbf{r} = \tfrac{1}{2}\mathbf{r} - \tfrac{1}{2}\mathbf{p} = \tfrac{1}{2}(\mathbf{r} - \mathbf{p})$$

Hence $MN = \tfrac{1}{2}QR$

Because MN is a multiple of QR, they must be parallel.

Test yourself

1 For a regular polygon of *N* sides, write down formulae for:

(a) the exterior angle   **(b)** the interior angle

2 Describe the minimum information you need to be told about two triangles to be sure that they are congruent to each other.

3 State which of the following pairs of triangles are congruent, giving a clear reason in each case.

(a)

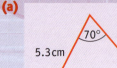

(b)

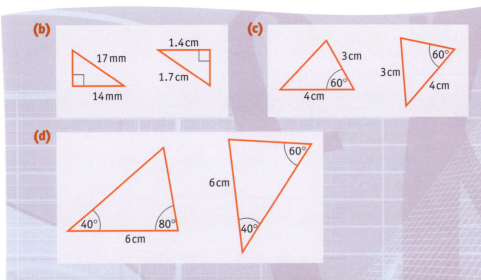

(c)

(d)

4 These circles have centre O. State the size of each angle labelled x, y or z.

(a)

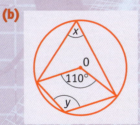

(b)

(c)

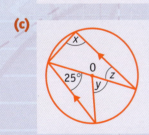

(d)

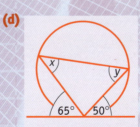

5 One interior angle of a hexagon is 140°. The other interior angles are all equal to one another. What are they?

6 P is a point on the major arc BC of a circle. The tangents at B and C meet at T. If ∠BPC = 65°, find ∠BTC.

7 Draw a rectangle 4 cm by 5 cm. Draw the locus of a point P that is 1 cm away from the perimeter of the rectangle.

8 (a) Draw a sketch of a square-based pyramid whose vertex is directly above the centre of the square.

(b) Describe the symmetries of the pyramid.

9 Calculate the size of the angles labelled x, y and z in these diagrams.

(a)

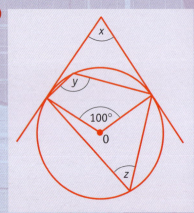

(b)

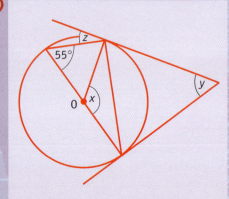

10

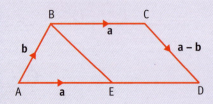

$AB = \mathbf{b}$, $AE = BC = \mathbf{a}$, and $CD = \mathbf{a} - \mathbf{b}$

(a) Find BE and AD in terms of $\mathbf{a}$ and $\mathbf{b}$.

(b) What type of quadrilateral is BCDE?

(c) What can be said about points A, E and D?

Practice questions

1

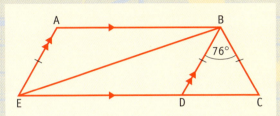

ABDE is a parallelogram, and point D is on the straight line EC. Work out the size of ∠BAE, giving a reason for your answer.

2 The diagram represents part of a regular decagon with two of its lines of symmetry shown.

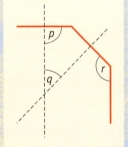

(a) Write down the value of angle *p*.

(b) Calculate the size of angles

(i) *q*

(ii) *r*

3

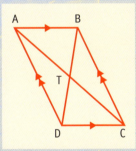

Which triangle is congruent to ABT? State why.

4 BT is a diameter of a circle and A and C are points on the circumference. The tangent to the circle at the point T meets AC produced at P (here, 'produced at' means 'continued until it meets'). Given that ∠ATB = 42° and ∠CAT = 26°, calculate:

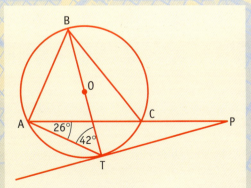

(a) ∠CBT   **(b)** ∠ABT   **(c)** ∠APT

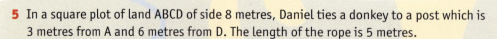

5 In a square plot of land ABCD of side 8 metres, Daniel ties a donkey to a post which is 3 metres from A and 6 metres from D. The length of the rope is 5 metres.

(a) Draw an accurate scale diagram to illustrate the region of the plot of land that the donkey can graze on.

(b) Use your diagram to calculate the area of the plot of land that the donkey cannot reach.

6 Each interior angle of a regular polygon is eight times each exterior angle. Calculate the sum of the interior angles of the polygon.

7

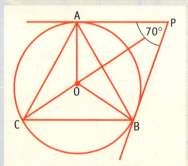

PA and PB are tangents to a circle centre O. POC is a straight line and ∠APB is 70°.

Calculate:

(a) ∠ABP

(b) ∠AOB

(c) ∠ACO

8 ABCD is a cyclic quadrilateral. AC is a diameter of the circle, centre O, which passes through A, B, C and D. ∠BAD = 68°, ∠DBC = 20°.

Calculate the size of:

(a) ∠BDA

(b) ∠AOD

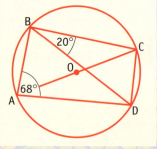

9

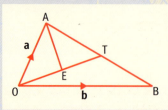

OA = **a** and OB = **b**. T is the mid-point of AB and E is the mid-point of OT. Find the value of OE in terms of **a** and **b**.

10 A man tries to swim straight across a river 25 metres wide, but the flow of the river takes him downstream. The man swims at a rate of 2.2 m/s in still water and the river is flowing at 1.5 m/s. How far downstream will the man end up after swimming across the river?

Statistics and probability

Frequency tables

When a lot of information has been gathered it is convenient to put it together in a frequency table. From this you can find the different averages (mean, median and mode).

The following table shows the number of people in cars on a stretch of motorway on a Saturday morning.

Number of people per car (N)	Frequency (f)	Number of people in these cars ($f \times N$)
1	72	$1 \times 72 = 72$
2	46	$2 \times 46 = 92$
3	38	$3 \times 38 = 114$
Total	**156**	**278**

The **mean** number of people in a car is found by adding the total number of people and dividing by the number of cars:

mean = $278 \div 156 = 1.782$

Hence the mean number of people per car is 1.78 (2 d.p.).

Grouped data

Often the information we are given is grouped together, as in the table below, which shows the range of pocket money given to children in a class.

Pocket money (£)	0.00–1.00	1.01–2.00	2.01–3.00	3.01–4.00	4.01–5.00
Number of children	2	7	12	3	1

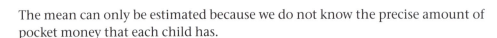

The mean can only be estimated because we do not know the precise amount of pocket money that each child has.

To estimate the mean we assume each person in each group has the 'mid-way' amount and we build up a table as shown below.

Pocket money (£)	Frequency (f)	Mid-way (m)	$f \times m$	Total (£)
0.00–1.00	2	0.50	2 × 0.50	1.00
1.01–2.00	7	1.505	7 × 1.505	10.535
2.01–3.00	12	2.505	12 × 2.505	30.06
3.01–4.00	3	3.505	3 × 3.505	10.515
4.01–5.00	1	4.505	1 × 4.505	4.505
Total	**25**			**56.615**

Note how we find the mid-way value: we add the upper and lower values in each category together and divide by two. We could round this off to the nearest penny if we wished since it is only an estimate, but it is usual not to do this rounding off until the last moment.

The estimated mean is £56.615 ÷ 25 = £2.26 (rounded off).

You will come across a few different ways of labelling the groups in a grouped frequency table.

For example, the initial table might have been labelled:

Pocket money (£x)	$0 \leq x \leq 1$	$1 < x \leq 2$	$2 < x \leq 3$	$3 < x \leq 4$	$4 < x \leq 5$

That is, in the group $2 < x \leq 3$, the pocket money is more than £2 but less than or equal to £3.

You will see different ways of using inequalities in this type of table. It makes no difference to the middle value as you still use the average of the end values.

Frequency polygons

Information given in a table such as the one below gives a simple frequency polygon.

Example 1

Number of patients	1	2	3	4	5
Frequency	10	19	20	17	6

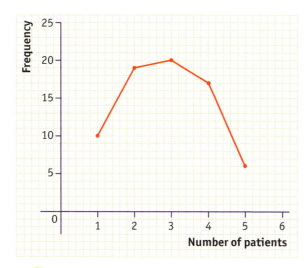

The frequency polygon is represented on a graph as shown.

Notice how we simply plot the coordinate from each ordered pair in the table ((1,10), (2,19) etc.).

Example 2

A doctor monitored how many visits she made to patients over 60 years old, over a period of 100 weeks. Below is a frequency table for the number of visits per week.

Number of visits made in 1 week	1–10	11–20	21–30	31–40	41–50
Frequency	4	21	35	29	11

The frequency polygon is shown.

Notice how we use the mid-point of each group just as we did in estimating the mean.

Notice also how we plot the ordered pairs of mid-points with frequency: (5.5, 4), (15.5, 21), (25.5, 35), (35.5, 29), (45.5, 11).

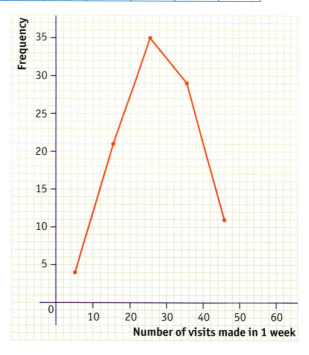

statistics and probability

Histograms

A histogram looks similar to a bar chart but with some fundamental differences:

- There are no gaps between the bars.
- The horizontal axis uses a continuous scale, since it is mainly used for continuous data such as time, weight or length.
- The *area* of each bar represents the frequency.
- The vertical axis is labelled 'frequency density', where
  frequency density = frequency ÷ width of interval

The table below shows the time taken to do a particular job, measured to the nearest minute.

Time (minutes)	1–2	3–7	8–10	11–12
Frequency	15	32	12	6

This information can be represented on a histogram:

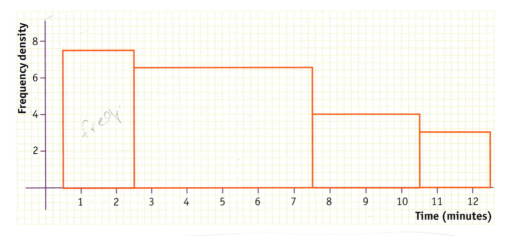

Notice that the histogram bars start and finish at the least possible and greatest possible times in each group.

For example, in the time region 3–7 minutes, the least possible time is $2\frac{1}{2}$ minutes and the greatest possible time is $7\frac{1}{2}$ minutes.

Take care when working out the height on the frequency density axis. You must remember that:

$$\text{height} = \frac{\text{frequency}}{\text{interval width}}$$

Follow through the chart below to see how the frequency densities were found from the data.

Time (minutes)	1–2	3–7	8–10	11–12
Interval width	(2.5 − 0.5) = 2	(7.5 − 2.5) = 5	(10.5 − 7.5) = 3	(12.5 − 10.5) = 2
Frequency	15	32	12	6
Frequency density	15 ÷ 2 = 7.5	32 ÷ 5 = 6.4	12 ÷ 3 = 4	6 ÷ 2 = 3

Surveys and questionnaires

A survey is an organised way of asking a lot of people a few well-constructed questions, or making a lot of observations in an experiment in order to reach a conclusion about something.

When designing a questionnaire, remember:

- never ask leading questions designed to get a particular response
- never ask irrelevant personal questions
- always keep the questions as simple as possible
- always set questions that will get a response from anyone who is asked
- giving boxes to tick narrows the range of possible answers

The following are examples of *bad questions*. Questions of these types should not appear on any of your questionnaires.

- What is your age? (This is personal, and many people will not want to answer.)
- Using poor defenceless animals to test cosmetics is cruel, don't you agree? (This is a leading question, designed to get a 'yes'.)
- Do you always fly when you go abroad? (This can only be answered by people who have been abroad.)
- If you are in town one day and you see a person asking for some money because he looks homeless, do you feel sorry for him and give him money or do you tell him where to get help from? (This is complicated — some might do neither.)

The following questions are good alternatives to those above.

- In which of the following age groups are you?

  0–20 ☐ 21–30 ☐ 31–50 ☐ Over 50 ☐

- Cosmetics should be tested on animals.

  Agree ☐ Disagree ☐ Don't know ☐

- If you went abroad, would you always fly?
- Would you give money to a beggar in the street?

A questionnaire is usually put together to test a hypothesis or a statement.

'Only people who live in Yorkshire like Yorkshire puddings with treacle on.'

Design a questionnaire that can be used to test whether this statement is true.

The questions we need to include are:

- Do you live in Yorkshire?
- Do you eat Yorkshire puddings?
- If you eat Yorkshire puddings, do you like treacle on them?

Once these questions have been answered, you can analyse the answers to see whether the statement is true or not.

Sampling

There are two main types of sample: random and stratified.

Random sampling

In a random sample, every member of the population has an equal chance of being selected — for example, every tenth person walking through the school gate, or names of 50 pupils drawn from a hat containing the names of all the pupils in the school, or the first 100 cars passing a certain road junction.

Stratified sampling

In a stratified sample, the population is divided into categories and the number to be chosen from each category is fixed at the start as the same ratio representing the whole population.

For example, say a school has the following distribution of pupils:

School year	Girls	Boys	Total
Y7	56	64	120
Y8	48	80	128
Y9	52	54	106
Y10	56	42	98
Y11	50	52	102
Total pupils in the school			554

To choose a stratified sample of 100 pupils from this school, we need to choose from each category as follows:

Y7 pupils — choose $\dfrac{120}{554} \times 100 = 22$ pupils (after rounding)

Of these 22 pupils, the number of Y7 girls should be $\dfrac{56}{120} \times 22 = 10$ (rounded)

The number of Y7 boys will then be 22 − 10 = 12

Similarly, for Y8 pupils, we need to choose $\frac{128}{554} \times 100 = 23$, of which $\frac{48}{128} \times 23 = 9$ will be girls and 14 boys.

This tells us how many of each sex from each year we must choose for our sample. Of course, these pupils then need to be chosen at random (unless there are other sub-divisions of category that we wish to include).

Sample size

The size of a sample is often determined by cost or time, but generally the larger the sample the more accurate is the representation of the true population.

Scatter diagrams

A scatter diagram is used to see if there is a connection between one variable and another. This connection is called a correlation.

Example

The scatter graphs below show the three different types of correlation.

1   Do taller people have bigger feet?

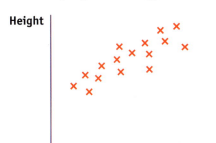

This diagram shows positive (or direct) correlation. It shows that the taller people are, the larger their feet are likely to be.

2   Is there a connection between temperature and the number of scarves sold?

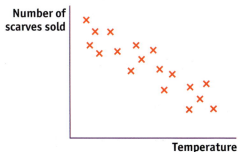

This diagram shows negative (or indirect) correlation. It shows that the higher the temperature, the fewer scarves you are likely to sell.

3 Is there a connection between weight of student and pocket money?

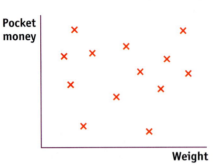

This diagram shows no correlation. It shows there is no correlation between weight and pocket money for students.

Lines of best fit

If we see there is a correlation between two variables then we can draw a line of best fit, that is, a line that follows the trend of the plotted data.

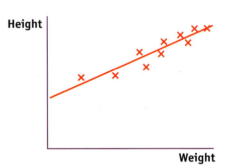

When you draw this line of best fit you should be trying to:

(i) show the trend

(ii) have about the same number of points above and below the line

(iii) draw the line from one side of the available graph to the other

But note that:
- this line *does not* have to go through all the points
- this line *does not* have to go through the origin
- this line is *not* drawn from the first to the last point
- this line is usually straight but it could be curved. However, until you get to A-level statistics, all the lines of best fit you will meet should be straight lines.

Cumulative frequency

This is commonly called the **running total**. It is used to draw a cumulative frequency diagram, from which we can estimate the median and find the quartiles.

Example

The following frequency table has a cumulative frequency column, which gives the running total.

Score	Frequency	Cumulative frequency
1–20	5	5
21–40	9	14
41–60	18	32
61–80	41	73
81–100	19	92

This information can be plotted as a cumulative frequency diagram (note that the points are plotted at the upper limit of the range). The points can be joined to give either a curve or a polygon. Both have been drawn for you below, so note the difference:

- the cumulative frequency curve has a smooth curve (called the ogive)
- the cumulative frequency polygon has straight lines between each point

Since the whole diagram is an approximation, it will make little difference which one you use, unless of course you are asked to use one particular type in an exam.

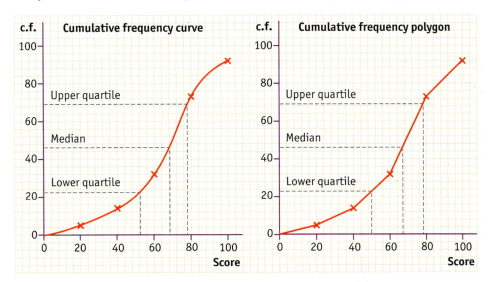

The readings you find from the two different types of diagram will be slightly different, but in the examination *your* diagram is the one that you are marked against.

Find the median by looking half-way along the cumulative frequency (the cumulative frequency is 92 in this case, so the median is at $92 \div 2 = 46$).

The quartiles come at the quarter marks: $\dfrac{92}{4} = 23$ and $92 \times \dfrac{3}{4} = 69$.

The interquartile range is the difference between the upper and the lower quartiles.

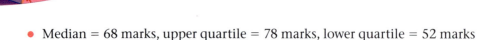

- Median = 68 marks, upper quartile = 78 marks, lower quartile = 52 marks
- The interquartile range = (78 − 52) = 26
- The semi-interquartile range = 13

Box and whisker diagrams

A box and whisker diagram displays data for comparison. This requires five data items:

(i) the lowest value

(ii) the highest value

(iii) the lower quartile

(iv) the upper quartile

(v) the median

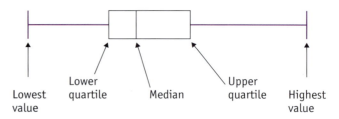

These data items are accurately plotted onto a scale.

The following diagram illustrates how the cumulative frequency and the box and whisker diagrams are connected to illustrate any distribution.

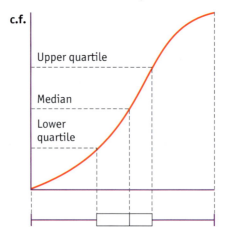

Example

The box and whisker diagram below shows the numbers of tomatoes on a truss for tomatoes grown with fertiliser X.

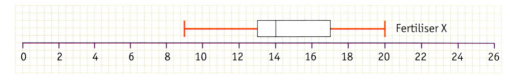

The results for the same type of tomatoes grown with fertiliser Y are: lowest number 5, highest number 22, lower quartile 11, upper quartile 16 and median 14.

Draw a box and whisker diagram for both types of fertiliser growth and comment on the differences.

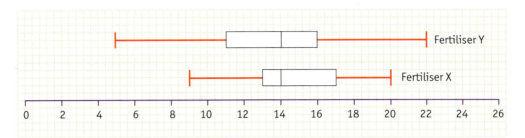

The growth for both types of fertiliser has the same median number of tomatoes on a truss, but both the lower and the upper quartiles for fertiliser Y are lower than those of fertiliser X. The range of tomatoes per truss, however, is larger with fertiliser Y compared with fertiliser X. Overall, the growth is better with fertiliser X, even though the truss with most tomatoes was grown with fertiliser Y.

Probability — a definition

The definition of probability is:

$$P(\text{event}) = \frac{\text{number of ways the event can happen}}{\text{total number of outcomes}}$$

This definition always leads to a fraction. If possible, the fraction should be cancelled down to its lowest terms. Sometimes you may be asked to give probability as a percentage or decimal.

Experimental probability

This can be found by performing an experiment many times or making an observation many times, and keeping an accurate record of the result.

The experimental probability of a particular event happening can then be worked out as:

$$\text{experimental probability} = \frac{\text{number of times the event happened}}{\text{total number of observations}}$$

Example

A normal dice was rolled 100 times. The number 2 was rolled a total of 21 times. This gives an experimental probability of:

$$\frac{21}{100} = 0.21$$

Theoretical probability

This is found by considering equally likely events. Equally likely events are those that all have an equal chance of happening.

For example, when rolling an unbiased dice, getting a 1, 2, 3, 4, 5 or 6 are equally likely events. In contrast, rolling two dice and getting totals of 2, 3, 4, 5, 6, 7 etc. are events that are *not* equally likely because, for example, there is only one way to score 2 but there are two ways to score 3 and three ways to score 4, and so on.

The theoretical probability of an event is found by the fraction:

$$\text{theoretical probability} = \frac{\text{number of ways the event can happen}}{\text{total number of different equally likely events that can occur}}$$

Examples

- Probability of rolling a dice and getting a 4 $= \frac{1}{6}$
- Probability of tossing a coin and getting a head $= \frac{1}{2}$
- Probability of a total of 4 when two dice are rolled $= \frac{3}{36} = \frac{1}{12}$

Expectation

The expectation of an event happening is found by multiplying the probability of the event by how many times the event has the opportunity of happening.

Example

I cut a pack of cards 200 times. How many times would I expect to get a king?

The expectation will be:

$$200 \times \frac{4}{52} = 15 \text{ (rounded)}$$

Probability of *not* happening

The probability of an event *not* happening is found by subtracting the probability of it happening from 1.

Example

- The probability of snow in March is 0.15.
- The probability of *no* snow in March is $1 - 0.15 = 0.85$.

The *or* rule

Two events are mutually exclusive when one can happen *or* the other but not both. If two events, A and B, are mutually exclusive then the probability of event A *or* B happening is equal to:

probability of A happening *added* to the probability of B happening

Example

Four brothers have a race.

- The probability of Kevin winning is 0.2
- The probability of Brian winning is 0.4
- The probability of Malcolm winning is 0.25
- The probability of David winning is 0.15

What is the probability of either Kevin or Malcolm winning?

The events are mutually exclusive (as only one brother can win), so we add their probabilities:

probability of Kevin or Malcolm winning = 0.2 + 0.25 = 0.45

Independent events

If two events can happen at the same time then we say that the events are independent.

Examples of independent events:

- Rolling two dice and getting a four and a two
- Tossing a coin twice and getting heads both times

The *and* rule

If we have two (or more) independent events happening, event A and event B, the probability of event A happening and then event B happening is found by:

multiplying the probability of event A by the probability of event B

Example

The probability of Amit getting his homework all correct is 0.15. The probability of Rachel getting her homework all correct is 0.65. What is the probability of them *both* getting the homework right?

Both implies the *and* rule, which means *multiply* the probabilities.

probability = 0.15 × 0.65 = 0.0975

Tree diagrams

A tree diagram can help us to see all the possibilities in a given situation and usually makes use of the *or* rule and the *and* rule in the same situation.

Follow through the following problem which uses a tree diagram.

Example

Two cards are dealt. What is the probability of:

(i) both being spades?

(ii) one being a spade?

(iii) at least one being a spade?

A tree diagram is drawn as shown below, with the probabilities on the branches. Note that the second set of probabilities depend on which card was chosen as the first card.

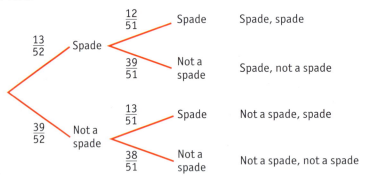

The probabilities of the combined events are calculated using the *and* rule.

(i) The probability of both cards being spades is:

$$\frac{13}{52} \times \frac{12}{51} = \frac{156}{2652} = 0.06 \text{ rounded}$$

(ii) The probability of one spade is either spade then not a spade:

$$\text{probability} = \frac{13}{52} \times \frac{39}{51} = \frac{507}{2652}$$

or not a spade then a spade:

$$\text{probability} = \frac{39}{52} \times \frac{13}{51} = \frac{507}{2652}$$

So the probability of one *or* the other happening is found by *adding* their probabilities:

$$= \frac{507}{2652} + \frac{507}{2652} = \frac{1014}{2652} = 0.38 \text{ (rounded)}$$

(iii) The probability of at least one being a spade can be found in two different ways:

- This happens when either you get one spade *or* two spades; hence *add* their probabilities:

$$\frac{1014}{2652} + \frac{156}{2652} = \frac{1170}{2652} = 0.44$$

- We could view this as the probability of *not* getting no spades at all; hence subtract the probability of not a spade *and* not a spade from 1, which is:

$$1 - \left(\frac{39}{52} \times \frac{38}{51}\right) = 1 - \frac{1482}{2652} = \frac{1170}{2652} = 0.44$$

It is useful to see both these methods because in some problems one method may be much better to use than the other.

Test yourself

1 Sketch a scatter diagram that has:

(a) positive correlation   **(b)** negative correlation   **(c)** no correlation

2 From the cumulative frequency diagram below, write down the following:

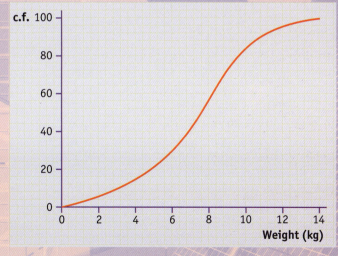

(a) median   **(b)** lower quartile

(c) upper quartile   **(d)** interquartile range

3 A bag contains four blue balls and eight white balls. What are the probabilities of choosing from the bag:

(a) a blue ball?

(b) two balls of the same colour?

(c) two balls of a different colour?

4 Construct a histogram to illustrate the following information:

Weight (kg)	1–2	3–5	6–10	11–13	14–20
Frequency	8	15	35	21	7

5 The following table shows some data on the marks for 50 boys and 50 girls in a mathematics exam:

	Lowest mark	Lower quartile	Median	Upper quartile	Highest mark
Boys	6	16	20	22	44
Girls	7	14	16	21	33

Draw a box and whisker diagram to illustrate these data.

Practice questions

1 All the Y11s took a spelling test. The table below shows the results.

Number of spellings correct (x)	Frequency
$0 \leq x < 20$	16
$20 \leq x < 30$	21
$30 \leq x < 40$	28
$40 \leq x < 50$	38
$50 \leq x < 60$	17
$60 \leq x < 100$	3
Total	**123**

(a) What is the modal class?

(b) Calculate the estimated mean number of correct spellings per pupil.

2 A marathon is held in Hope one year.

(a)

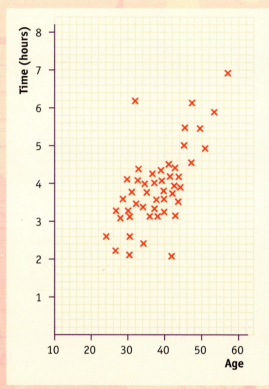

The scatter diagram shows the ages of runners and the times they took.

What does the scatter diagram tell you?

(b) The table shows the times taken by 250 men and 250 women:

Time taken (*t* hours)	Men	Women
$2 < t \leq 3$	44	16
$3 < t \leq 4$	91	34
$4 < t \leq 5$	58	82
$5 < t \leq 6$	32	78
$6 < t \leq 7$	18	21
$7 < t \leq 8$	7	19

Draw a frequency polygon to show the distribution of times taken.

(c) The frequency polygon for the women's times is different from that for the men's times.

Describe how they differ.

3 Helen and James played 40 games of golf together. The table below shows Helen's scores.

Scores (x)	$70 < x \leq 80$	$80 < t \leq 90$	$90 < t \leq 100$	$100 < t \leq 110$	$110 < t \leq 120$
Frequency	2	5	14	16	3

(a) Draw a cumulative frequency diagram to show Helen's scores.

(b) Use your graph to find:

    (i)   Helen's median score

    (ii) the interquartile range of her scores

(c) James's median score was 102. The interquartile range of his score was 7.

    (i)   Who was the more consistent player? Give a reason for your choice.

    (ii) The winner of a game of golf is the one with the lowest score. Who won most of these 40 games? Give a reason for your choice.

4 A sample of 50 pupils were asked how long it took them to get to school. Their replies were recorded, to the nearest minute, as:

Time (minutes)	9–10	11–15	16–20	21–30
Number of pupils	2	6	10	32

Draw a histogram to display these data.

5 This table gives the weekly wages of 200 employees in a local firm.

Wage (£)	≤ 150	151–200	201–300	301–400	401–500	501–600	601–800
Number of employees	20	15	26	71	29	21	18

(a) Construct a cumulative frequency diagram for this distribution.

(b) What is the median wage?

(c) What is the interquartile range?

(d) How many workers earn more than £700 per week?

6 The age distribution of males aged under 90 in the UK (given in percentage) is shown in the following table.

Age	England	Scotland	Wales	Northern Ireland
0–9	12.9	12.7	12.9	16.7
10–19	12.9	13.4	13.1	16.4
20–29	16.2	16.5	15.7	16.4
30–39	13.6	13.8	12.8	12.9
40–49	13.0	12.5	12.7	11.5
50–59	10.6	11.0	10.7	9.3
60–69	10.3	10.2	11.0	8.2
70–79	6.9	6.6	7.3	5.7
80–89	3.6	3.3	3.8	2.9

(a) (i) On the same graph, draw cumulative frequency diagrams for the age distributions in England and Northern Ireland.

(ii) State the interquartile range of the age distribution for England and for Northern Ireland.

(iii) What differences can you state about the age distributions of England and Northern Ireland from the graph?

(b) What differences can you state about the age distributions of England and Scotland?

(c) Make some observations about the age distribution throughout the UK.

7 In the Klingon Homeworld, out of every 100 children born there are usually 35 females. What is the probability of a family with three children having:

(a) all three girls?   **(b)** two girls and one boy?   **(c)** at least one girl?

8 The probability that it rains in April during the busy morning traffic is 0.4. The probability of bad hold-ups in the traffic during the busy mornings is 0.33 during rain and 0.25 when there is no rain.

(a) Calculate the probability that whatever the weather, there will be a bad traffic hold-up on the morning of 26 April.

(b) On approximately how many mornings in an average April will there be bad hold-ups in the traffic? (Assume there will be 25 working days in April and hence 25 days of busy morning traffic.)

9 (a) Design a questionnaire to test the theory that for motorists with over 5 years' driving experience, men are more likely to have a car accident than women.

(b) State clearly how you would choose your sample to complete this questionnaire.

Answers and hints

Number

Test yourself

1. **(a)** 5.85 kg   **(b)** 3.95 m

2. 10.8 kg (3 s.f.)

3. **(a)** 800   **(b)** 80   **(c)** 1200   **(d)** 160

4. £130:£195

5. £115

6. **(a)** $1\frac{4}{15}$ **(b)** $\frac{3}{40}$ **(c)** $\frac{2}{3}$ **(d)** $1\frac{3}{7}$

7. **(a)** $3.5 \leq 4\,\text{cm} < 4.5$ **(b)** $1.945 \leq 1.95\,\text{m} < 1.955$ **(c)** $7.225 \leq 7.23\,\text{kg} < 7.235$

8. **(a)** 9 and –9   **(b)** 3   **(c)** 8   **(d)** 125   **(e)** 10000

9. **(a)** 90   **(b)** $60\sqrt{2}$ **(c)** $72\sqrt{7}$

10. There are many possibilities but some of the more obvious ones are:

    (a) $\pi + 2$, $\sqrt{30}$ or $4 + \sqrt{2}$ **(b)** $50 + \pi$, $\sqrt{3000}$ or $60 - \sqrt{2}$

11. **(a)** 5.684×10^3 **(b)** 5.0×10^6 **(c)** 3.82×10^4 **(d)** 9.5×10^{-3}

12. **(a)** 8.5×10^{12} **(b)** 7.0×10^{-5}

13. **(a)** 1.23×10^{-11} **(b)** 4.76×10^{-12}

Practice questions

1. $\dfrac{48 \times 0.5}{6} = 8 \times 0.5 = 4$

Exam tip: Always round off to something suitable. Round 46.7 to 48 so that you can divide by 6 (6.23 rounded). Round 0.44 either to 0.4, to give an estimate of 3.2, or to 0.5, to give an estimate of 4.

2   The area of the triangle is found by multiplying half the base length by the height. The upper bound of the area will be calculated using the upper bounds of the measurements given, i.e. height 8.65 cm and base length 6.35 cm.

upper bound of area $= \frac{1}{2} \times 6.35 \times 8.65 = 27.46375$ or 27.5 cm^2

Similarly, the lower bound is found by using the lower bounds of the given information, i.e. height 8.55 cm and base length 6.25 cm, so giving the area as:

$\frac{1}{2} \times 6.25 \times 8.55 = 26.71875$ or 26.7 cm^2

The upper and lower bounds for the area of the triangle are 27.5 cm^2 and 26.7 cm^2.

3   Malcolm: £3 456 576 $\times \dfrac{4}{28} =$ £493 796.57

Janet: £3 456 576 $\times \dfrac{6}{28} =$ £740 694.86

Neil: £3 456 576 $\times \dfrac{18}{28} =$ £2 222 084.57

It would not be correct to round off these figures in this particular context.

4   **(a)** distance = speed × time. 45 (2 s.f.) is between 44.5 and 45.5.

(i) The upper bound will use the largest values, i.e. 45.5 and 8.5. We need to change 45.5 miles per hour into miles per minute:

= 45.5 ÷ 60 = 0.758333 miles per minute (don't round off yet)

distance = 0.758333 × 8.5 = 6.45 miles (3 s.f.)

(ii) The lower bound will use the smallest values, i.e. 44.5 and 7.5. We need to change 44.5 miles per hour into miles per minute:

= 44.5 ÷ 60 = 0.7416666 miles per minute (don't round off yet)

distance = 0.7416666 × 7.5 = 5.56 miles (3 s.f.)

(b) Since both figures round off to 6 miles, this would be the most sensible figure to give as the distance between the school and the library.

5 $\frac{19}{32}$ inches

You need to add the two fractions together by first changing $\frac{7}{8}$ to $\frac{14}{16}$. They add to give $\frac{19}{16}$. This then needs to be halved, i.e. divide by 2. This is best done by doubling the denominator (2 × 16 = 32).

6   **(a)** £5755

Use the rule 5000 × 1.048 × 1.048 × 1.048 or 5000 × 1.048^3.

(b) 1.157625

This is derived from (1.05)^3.

7 (a) 4 (the cube root of 64)

(b) $16^{\frac{1}{2}} = (2^4)^{\frac{1}{2}} = 2^2$ and $8^{-3} = (2^3)^{-3} = 2^{-9}$, so $16^{\frac{1}{2}} \times 8^{-3} = 2^2 \times 2^{-9} = 2^{-7}$

(c) $125^{\frac{1}{3}} = 5$, hence $y = \frac{1}{3}$

Exam tip: You will get credit for knowing what negative indices mean and what fractional indices mean, so show your working in case you make a careless error in the calculation.

8 (a) $5 + \sqrt{3}$ and π^2 are both irrational, so the only rational answer is part (iii).

(b) There are many possible answers. Two examples are:
$$3\sqrt{2} \times \sqrt{2} = 3 \times 2 = 6 \text{ and } 3\pi \times \frac{1}{\pi} = 3$$

9 (a) $(\sqrt{3} + \sqrt{12})^2 = \sqrt{3} \times \sqrt{3} + \sqrt{3} \times \sqrt{12} + \sqrt{12} \times \sqrt{3} + \sqrt{12} \times \sqrt{12}$
$$= 3 + \sqrt{36} + \sqrt{36} + 12$$
$$= 3 + 6 + 6 + 12 = 27$$

It is important to show all the steps that lead you from the bracket, through the expansion, to the final 27.

(b) (i) $\dfrac{\sqrt{18}}{\sqrt{6}} = \dfrac{\sqrt{6} \times \sqrt{3}}{\sqrt{6}} = \sqrt{3}$

(ii) $\sqrt{6} \times \sqrt{18} \times \sqrt{27} = \sqrt{(6 \times 18 \times 27)} = \sqrt{(6 \times 6 \times 3 \times 3 \times 9)}$
$$= 6 \times 3 \times 3 = 54$$

It is important to show how you have joined surds together or simplified them, especially on the non-calculator paper.

10 $\sqrt{18} + \sqrt{72} = \sqrt{18} + \sqrt{(4 \times 18)}$
$$= \sqrt{18} + \sqrt{4} \times \sqrt{18} = \sqrt{18} + 2 \times \sqrt{18} = 3\sqrt{18}$$
$$= 3 \times \sqrt{(9 \times 2)} = 3 \times \sqrt{9} \times \sqrt{2}$$
$$= 3 \times 3 \times \sqrt{2} = 9\sqrt{2}$$

Exam tip: The final expression needs to be as simple as possible, which is why the best answer gets this down to $9\sqrt{2}$ rather than leaving it as $3\sqrt{18}$, which would probably earn just 1 mark out of 2.

11 (a) 54 000 000

(b) $2.5 \times 10^9 \div 5.4 \times 10^7 = £46.30$ (£46 would be accepted)

Exam tip: In part **(b)** it is important to recognise a billion as a thousand million, then do the calculation in standard form on the calculator.

12 **(a)** $20\,000 \times 365 = 7\,300\,000 = 7.3 \times 10^6$ tonnes

(b) $\dfrac{5.29 \times 10^4}{7.3 \times 10^6} \times 100 = 0.72\%$

Exam tip: Setting up the calculation here would not gain you marks. It is being able to show that you can actually work in standard form and give the answer in an appropriate form that earns the marks.

Algebra

Test yourself

1 **(a)** $x = 1 \quad y = -5$ **(b)** $x = 1.5 \quad y = 2.5$

2 You should find cost $= 7.5\sqrt{\text{(number of people)}}$
     (a) £106.07   **(b)** 169

3 **(a)** $8x^2 - 10x - 3$ **(b)** $3x^2 - 4x - 15$ **(c)** $2mp + 2pt - 3mt$

4 **(a)** $x = 7$ **(b)** $x = -1$ **(c)** $x = 2.375$

5 **(a)** $y = \dfrac{x + 2}{2}$ **(b)** $y = \dfrac{x}{b + 7}$ **(c)** $y = \dfrac{7t - p}{5}$

6 **(a)** $t(3 + 7t)$ **(b)** $2m^2(m - 3)$ **(c)** $3mp(2p^2 + 3mt)$
    (d) $(x - 3)(x - 4)$ **(e)** $(x + 5)(x - 5)$ **(f)** $(2x + 5)(x - 3)$

7 **(a)** $x = -3$ and $x = -4$ **(b)** $x = 0.27$ and $x = -1.47$

8 **(a)** $x > 6.4$ **(b)** $t > 8$ **(c)** $-6 < x < 6$ **(d)** $-1 \leq x < 0.2$

9 **(a)** $5n - 3$ **(b)** $n^2 + 2$ **(c)** $\dfrac{3n - 2}{4n + 1}$

Practice questions

1 **(a)** $7.5\,(0.35 + 25) = 190.125$ **(b)** $(56.25 - 0.1225) \div -5 = -11.2255$

Exam tip: There is no reason to round off here, although it is unlikely to lose you any marks provided your original unrounded off answer is shown.

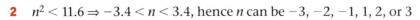

2 $n^2 < 11.6 \Rightarrow -3.4 < n < 3.4$, hence n can be $-3, -2, -1, 1, 2$, or 3

3 **(a)** $x = 8.6$ **(b)** $x = -0.55$ **(c)** $x = \dfrac{6}{17}$

Exam tip: The answer to part **(c)** is not exact unless you give $\frac{6}{17}$. If you give the answer as a rounded off decimal, then you could lose a mark.

4 **(a)** $x^2 - a^2$

(b) You should recognise the difference between two squares even without part **(a)**. Substitute into the expression to get $(978.5 + 21.5)(978.5 - 21.5)$ to get $1000 \times 957 = 957\,000$. You should be able to check this on your calculator.

5 Multiply the top equation by 3 and the bottom equation by 2 to enable y to be eliminated by adding the two equations. This will lead to the solution $x = 3.5$ and $y = -1.5$.

6 **(a)** (i) $n > 11.8$ (ii) 12

(b) $-1.2 < x < 1.2$

7 The constant of proportionality is not a whole number, so evaluate it and put it into your calculator's memory. It should be about 0.2153061.

(a) cost $= k \times 9^2 = £17.44$ (rounded)

(b) radius $= \sqrt{\dfrac{26.05}{k}} = 11\,\text{mm}$ (rounded)

8 **(a)** $12x^2 + 7x - 10$ **(b)** $4x(2 - 3x)$

(c) $M = \dfrac{9 - (7T)^2}{5}$ **(d)** $x^{-4}y^{-2}$ so $a = -4$ and $b = -2$

9 **(a)** (i) To get the $x^2 + 6x$ part you need $(x + 3)^2 = x^2 + 6x + 9$. Add another 2 to get $x^2 + 6x + 11$, so $x^2 + 6x + 11 = (x + 3)^2 + 2$.

(ii) $(x + 3)^2$ can be no lower than 0, since it cannot be negative. So the lowest value of $x^2 + 6x + 11$ is the same as for $(x + 3)^2 + 2$, which is 2.

(b) The phrase 'to 2 decimal places' should tell you to use the formula, where $a = 4$, $b = 6$ and $c = -11$. This gives the solutions $x = 1.07$ and $x = -2.57$.

10 **(a)** (i) $\dfrac{24}{25}$ (ii) $\dfrac{2n}{(2n + 1)}$ **(b)** $\dfrac{3}{2}, \dfrac{5}{4}, \dfrac{7}{6}\ldots$

GCSE Mathematics Revision Guide

Graphs

Test yourself

1 The point of intersection is at (0.7, 2). So the solution is $x = 0.7$ and $y = 2$.

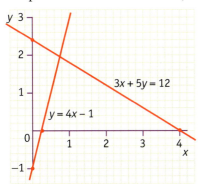

2 **(a)**

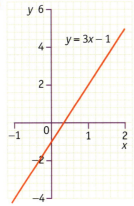

(b)

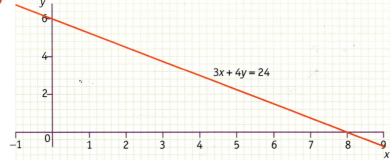

(c)

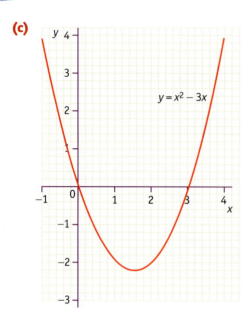

$y = x^2 - 3x$

(d)

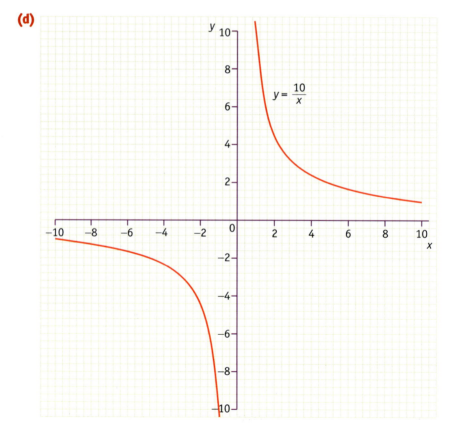

$y = \dfrac{10}{x}$

(e)

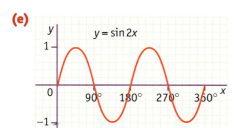

(f)

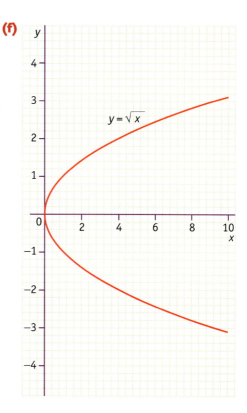

3

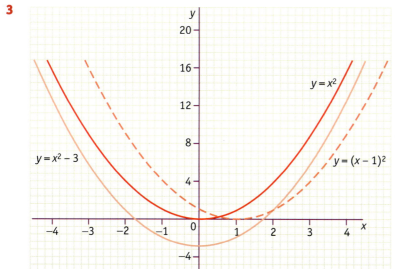

4 **(a)** 3   **(b)** 2   **(c)** $-\frac{1}{3}$

5 **(a)** speed   **(b)** acceleration   **(c)** distance travelled

6

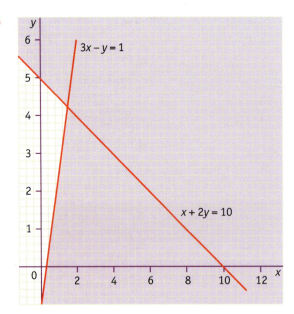

The maximum integer value of $x + y$ is 5 (from both (0, 5) and (1, 4)).

7 (a)

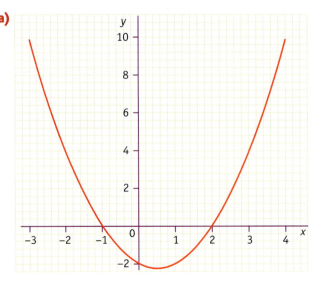

(b) $x = 2.3$ and -1.3 **(c)** $x = 2.6$ and -1.6

Practice questions

1 (a)

Number of weeks (w)	0	1	2	3	4	5
Amount of waste left (T)	200	120	72	43.2	25.9	15.5

(b)

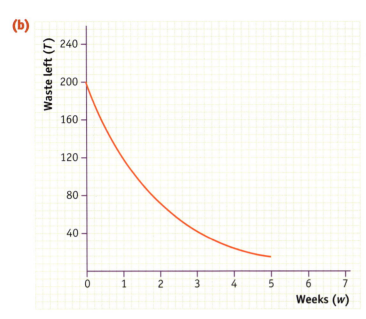

(c) If you extend the graph, you will find that it goes below 10 tonnes after 6 weeks.

2 (a) Find the steepest gradient on the graph. This is at the very beginning, 64 m/s.

(b) 1100 metres in 60 seconds is 18.3 m/s.

(c) You need to split up the area under the curve into trapeziums and find the area (about 3900 metres).

3 (a)

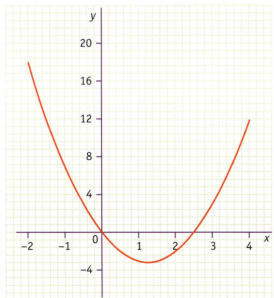

(b) Draw a line of gradient 4 beside your graph, then look for a parallel gradient. You will find such a gradient at the point $(2.2, -1.1)$.

4 You should draw two graphs, $y = \frac{1}{x}$ and $y = x^2 - 2$. Your solutions will be any points of intersection.

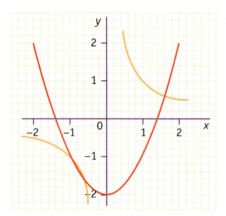

You should find three points of intersection, at $(-1, -1)$, $(-0.6, -1.6)$ and $(1.6, 0.6)$. It is only the x value that we need to solve the equation. Hence the solutions are $x = -1$, $x = -0.6$ and $x = 1.6$.

5 (a)

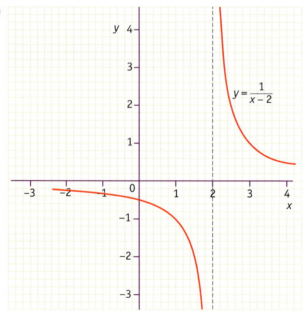

(b)

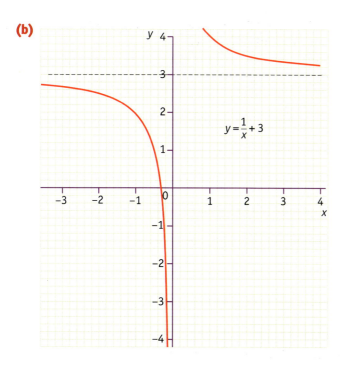

$$y = \frac{1}{x} + 3$$

(c)

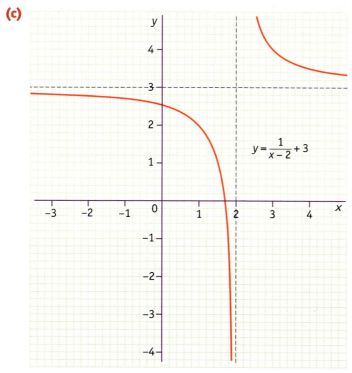

$$y = \frac{1}{x-2} + 3$$

6 (a)

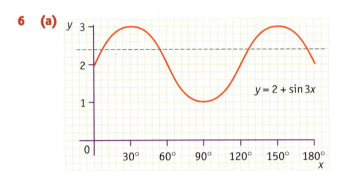

$y = 2 + \sin 3x$

(b) We need to find where $2 + \sin 3x = 2.4$ (this is $\sin 3x = 0.4$). The line on the graph shows the points where $2 + \sin 3x = 2.4$. This gives values of x at approximately 8°, 52°, 128° and 172°.

7 (a)

Time (hours)	0	3	6	9	12	15	18
Number of cells	1	2	4	8	16	32	64

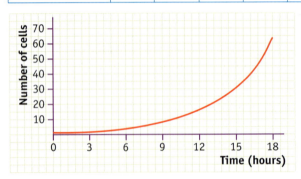

(b) $C = 2^{T/3}$

(c) We are looking for the solution to $1000 = 2^{T/3}$. We can find that $2^{10} = 1024$, which puts $T/3 = 10$, i.e. $T = 30$. So the bacteria will be dangerous after just 30 hours.

8 (a) The differences are reducing each time, which means it cannot be graph A or B, and R increases as S does, so it cannot be graph C. This only leaves graph D.

(b) You should recognise the shape as $y = \sqrt{x}$. $R = k\sqrt{S}$

(c) Choose a point to substitute into the equation. $S = 25$ and $R = 11.2$ would be the best since it gives an exact square root.

$$11.2 = k\sqrt{25} = 5k \Rightarrow k = 2.24$$

9 (a) The gradient is 3, so the line is $y = 3x + c$. Substitute $x = 0$ when $y = 5$ to find $5 = c$. Hence, the equation of the line is $y = 3x + 5$.

(b) Similarly, you should find the equation of the line is $y = 3x - 1$.

10 (a) You should draw a set of linear inequalities and find the feasible region.

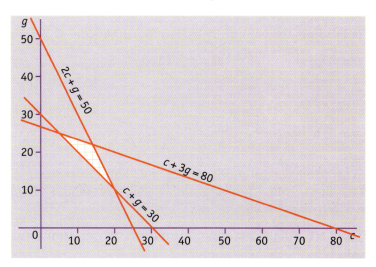

Then you can investigate the few points in that region. The lines to draw are $500c + 1500g = 40000$, which simplifies to $c + 3g = 80$; and $c + g = 30$ with $2c + g = 50$. This gives you a few points to look at. You should find the optimum number of machines is 5 cutters and 25 grinders, giving a daily output of 1500 widgets.

(b) The cost is $(5 \times 500) + (25 \times 1500) = £40\,000$.

Shape

Test yourself

1 **(a)** $432\,\text{cm}^3$ **(b)** $445\,\text{cm}^3$

2 **(a)** $378\,\text{g}$ **(b)** $2.42\,\text{g/cm}^3$

3 **(a)** arc length $= 1.9\,\text{cm}$, sector area $= 5.7\,\text{cm}^2$

    (b) Find the angle from the arc length information. This is $40.1°$, but use the calculator value to substitute into the area formula to give $8.75\,\text{cm}^2$.

4 **(a)** $V = 419\,\text{cm}^3$, $CSA = 263\,\text{cm}^2$ (all to 3 s.f.)

    (b) $V = 518\,\text{cm}^3$, $CSA = 283\,\text{cm}^2$ (all to 3 s.f.)

    (c) $V = 12\,100\,\text{cm}^3$, $CSA = 2230\,\text{cm}^2$ (all to 3 s.f.)

5 **(a)** $60\,\text{cm}^3$ **(b)** $14\,\text{cm}^3$

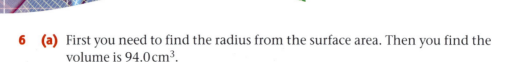

6 **(a)** First you need to find the radius from the surface area. Then you find the volume is $94.0\,cm^3$.

(b) Find the radius from the volume; then find the surface area is $192\,cm^2$.

7 **(a)** $x = 10.8$ and $y = 5.6$ **(b)** $x = 4.2$ and $y = 5$

Practice questions

1 volume $= \frac{1}{3}$ base area $\times$ height $= \frac{1}{3} \times 160 \times 18 = 960\,cm^3$

2 Calculate the curved area of the cone to be about $233\,cm^2$. Use this curved area to find the radius of the sphere from the surface area; this is about $4.3\,cm$. Then use the radius to calculate the volume as $336\,cm^3$.

Exam tip: Do not round off until the last stage — always use the calculator answer from one stage to the next.

3 Calculate the difference in volume of the two spheres to find the volume of the shell. Multiply by the density to give a weight of $9382\,grams$, which is $9.38\,kg$.

4 The side is a trapezium, so the whole shape is a prism with a regular cross-section of a trapezium. The volume is given by trapezium area $\times 12\,m = 80 \times 12 = 960\,m^3$.

5 Substitute the given values into the formula for the volume of a cylinder and then rearrange to get height $= 9.95\,cm$.

6 Find the volume of the cylinder, and the volume of the space left in the pan. The cylinder has a volume of $869\,cm^3$, the volume left is only $628\,cm^3$, so, yes, the water level will rise above the pan.

7 $x = 2.75\,cm$, $y = 16\,cm$

8 **(a)** (i) $20\,cm$

(ii) $7\,cm$

(b) Find the length ratio by cube-rooting the volume ratio, which will be the same as the weight ratio since they are directly proportional. Then calculate the large jar height as $12.9\,cm$.

9 Use the arc length of the sector to find the circumference of the base of the cone. Then find the radius of the base of the cone (using $2\pi r$). Find the height of the cone from the slant height $10\,cm$ and the radius using Pythagoras's theorem. You can then use the formula to calculate the volume of the cone as $289\,cm^3$.

10 Use similar triangles to find the height of the 'larger cone' (18.57 cm) and thus the height of the 'top' cut-off cone (3.57 cm). Then find the slant heights of each cone using Pythagoras's theorem (31.95 cm for the large cone, 6.14 cm for the cut-off cone) in order to find the curved surface areas. The difference between these is the surface area of the shade, which is 2520 cm² (3 s.f.).

Pythagoras and trigonometry

Test yourself

1 **(a)** 4.6 cm   **(b)** 3.8 cm   **(c)** 7.4 cm

   (d) 7.6 cm   **(e)** 3.6 cm   **(f)** 8.0 cm

2 **(a)** 39.8°   **(b)** 41.8°   **(c)** 45.6°

3 **(a)** 7.3 cm   **(b)** 8.7 cm   **(c)** 8.6 cm

4 **(a)** 7.9 cm   **(b)** 101°   **(c)** 134°

5 **(a)** $x = 9.4$ cm, $y = 48.5°$ **(b)** $x = 23°$, $y = 7.7$ cm   **(c)** $x = 7.1$ cm, $y = 83.9°$

Practice questions

1 $AB^2 = 6^2 + (5.5 - 4)^2 = 38.25$
$AB = \sqrt{38.25} = 6.18$ m

2 $BD = \dfrac{28}{\tan 41°} = 32.21$ m, $BC = \dfrac{28}{\tan 53°} = 21.099$ m

$CD = BD - BC = 11.1$ m

3 **(a)** Use the cosine rule on triangle ABC to find the angle at B. This will give the angle at A, and you can use the cosine rule again to find the length BD as 13.3 cm.

(b) The diagonals will bisect each other. Hence we have a triangle with sides 3, 9 and 6.65. The obtuse angle is opposite the 9 cm side and found by the cosine rule:

$$\cos\theta = \frac{3^2 + 6.65^2 - 9^2}{2 \times 3 \times 6.5} = -0.6961779$$

$$\theta = 134°$$

4 Use Pythagoras's theorem to find the diagonal of the base of the box, then the diagonal of the whole box. This gives 30.2 cm.

5   Use the sine rule to set up $\sin B = \dfrac{27 \sin 53°}{34}$.

The smaller the angle B, the smaller is the sine of B, so we want to know the smallest possible value of $\sin B$. Use the lower bounds of 27 and 53, with the upper bound of 34. This gives:

$$\sin B = \frac{26.5 \sin 52.5°}{34.5} = 0.6093873 \Rightarrow B = 37.5°$$

6   The key to this question is drawing a good diagram.

Use the cosine rule to find the length of the return trip as 12.4 km. Use the sine rule to calculate angles to find the bearing is 280°.

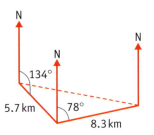

7   If the area of half the parallelogram $= \frac{1}{2} \times AB \times BC \times \sin B$, then the area of the whole parallelogram will be:

$$= AB \times BC \times \sin B$$
$$= 4 \times 3 \times \sin 149 = 6.2\,\text{cm}^2$$

8   Again, the key is a good diagram.

Use the sine rule to find the distance (104 m) Sam was away from the top of the building at first. Then use right-angled trigonometry to find the height of the building as 47.2 m.

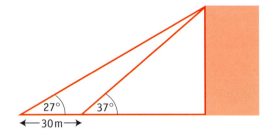

9   **(a)** Write down the formula for the arc length, add $2r$ to give the total perimeter of 20 cm, then rearrange the formula to evaluate r to be 6.68 cm.

   (b) Find the area of the sector, then multiply by the thickness (don't forget to change the units) to get a volume of 6.66 cm³.

10   **(a)** Use the area formula with sine to find the angle at A is 42°.

   (b) Use the cosine rule to find the length of BC as 6.249 cm, then use the sine rule to find the angle at B is 61.4°.

Geometry and construction

Test yourself

1 **(a)** $\dfrac{360}{N}$ **(b)** $180 - \dfrac{360}{N}$

2 All three sides correspond; two sides correspond with the same included angle; all angles correspond with one corresponding side; or a right angle with any two other corresponding sides.

3 **(a)** congruent, SAS **(b)** congruent, RHS
   (c) not congruent **(d)** congruent, ASA

4 **(a)** $x = 45°$ **(b)** $x = 55°, y = 125°$
   (c) $x = 90°, y = 50°$ and $z = 25°$ **(d)** $x = 50°, y = 65°$

5 $116°$

6 $50°$

7

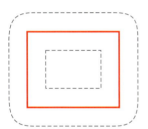

8 **(a)**

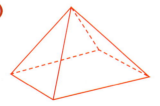

(b) It has four planes of symmetry. Planes of symmetry bisect the square base:
   (i)   front to back
   (ii) side to side
   (iii) diagonally left back to front right
   (iv) diagonally right back to left front
   It also has one axis of rotational symmetry through the apex and the centre of the square base of order 4.

9 **(a)** $x = 80°, y = 130°, z = 50°$ **(b)** $x = 110°, y = 70°, z = 35°$

10 (a) BE = **a** − **b**; AD = 2**a**

(b) A parallelogram

(c) They are all in a straight line (collinear).

Practice questions

1 ∠BDC = $\frac{1}{2}$(180 − 76) = 52°, ∠BDE = 180 − 52 = 128° = ∠BAE

2 (a) 90°   **(b)** (i) 36° $\left(\dfrac{360}{10}\right)$ (ii) 144° $\left(180 − \dfrac{360}{10}\right)$

3 ABT is congruent to CDT since ∠DCA = ∠BAC (alternate angles) and ∠ABD = ∠CDB (alternate angles). AB = DC because ABCD is a parallelogram, so we use rule ASA to show they are congruent.

4 (a) 26° (angles in the same sector)

(b) 48° (angles in a right angle)

(c) 22° (angles in a triangle)

5 (a) Find the post by drawing an arc 3 cm from A and another 6 cm from D. The post is at the point where they cross.

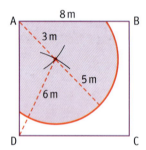

(b) Split your diagram into triangles and a sector to estimate the land used.

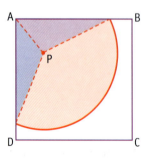

• First, find the two angles at corner A:

$$\cos\theta = \frac{8^2 + 3^2 - 6^2}{2 \times 8 \times 3} = 0.7708$$

$$\theta = 39.57$$

- Calculate other necessary angles in sequence shown below

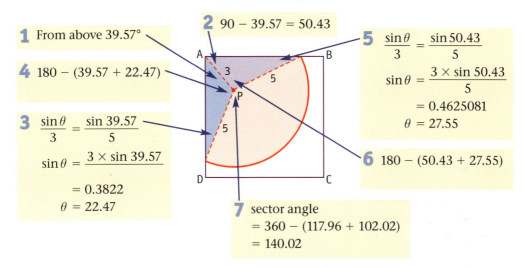

1 From above 39.57°

2 90 − 39.57 = 50.43

4 180 − (39.57 + 22.47)

5
$$\frac{\sin\theta}{3} = \frac{\sin 50.43}{5}$$
$$\sin\theta = \frac{3 \times \sin 50.43}{5}$$
$$= 0.4625081$$
$$\theta = 27.55$$

3
$$\frac{\sin\theta}{3} = \frac{\sin 39.57}{5}$$
$$\sin\theta = \frac{3 \times \sin 39.57}{5}$$
$$= 0.3822$$
$$\theta = 22.47$$

6 180 − (50.43 + 27.55)

7 sector angle
= 360 − (117.96 + 102.02)
= 140.02

- The areas of the triangles are found by $\frac{1}{2}ab\sin\theta$
$$(\tfrac{1}{2} \times 3 \times 5 \times \sin 117.96) + (\tfrac{1}{2} \times 3 \times 5 \times \sin 102.02)$$
$$6.6245635 \quad + \quad 7.3355622 \quad = 13.96 \ (2 \text{ d.p.})$$
- The area of the sector is found by
$$\frac{140.02}{360} \times \pi \times 5^2 = 30.55 \ (2 \text{ d.p.})$$
- Add the triangles to the sector to give 44.51
- So the area of land the donkey cannot reach is $8^2 − 44.51 = 19.49$, which is $19.5\,\text{m}^2$ (1 d.p.)

6 Set up an equation between interior and exterior angles:

$$\text{exterior angle} = \frac{360°}{n}$$

$$\text{interior angle} = 180° - \frac{360°}{n}$$

Hence $\dfrac{360 \times 8}{n} = 180 - \dfrac{360}{n}$

i.e. $\dfrac{360 \times 8}{n} = \dfrac{180n - 360}{n}$

∴ $2880 = 180n - 360$

$3240 = 180n$

$n = 18$

The polygon has 18 sides. Hence multiply the interior angle (160°) by 18 to get 2880°.

7 **(a)** 55°

    (b) 110°

    (c) 27.5°

8 **(a)** 42°

    (b) 140°

9 $\frac{1}{4}(\mathbf{a} + \mathbf{b})$

10 A good diagram will suggest using vectors as forces, or similar triangles. Either way, you should find an answer of 17 metres (2 s.f.).

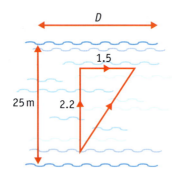

$$\frac{D}{25} = \frac{1.5}{2.2}$$

$$D = \frac{25 \times 1.5}{2.2} = 17\,\text{m}$$

Statistics and probability

Test yourself

1 **(a)**    **(b)**    **(c)**

2 **(a)** 7.6 kg   **(b)** 5.4 kg   **(c)** 9.1 kg   **(d)** 3.7 kg

3 (a) $\frac{4}{12}$, which cancels down to $\frac{1}{3}$

(b) You need to find the probability of two blues, two whites, then add them together... $\left(\frac{4}{12} \times \frac{3}{11}\right) + \left(\frac{8}{12} \times \frac{7}{11}\right) = \frac{17}{33}$ (0.515)

(c) This will be the rest of the possibilities from part **(b)**, hence

$1 - \textbf{(b)} = \frac{16}{33}$ (0.485).

4 Find the heights on the frequency density by dividing the frequency by the interval width.

Weight (kg)	1–2	3–5	6–10	11–13	14–20
Width	2	3	5	3	7
Frequency	8	15	35	21	7
f.d.	8 ÷ 2 = 4	15 ÷ 3 = 5	35 ÷ 5 = 7	21 ÷ 3 = 7	7 ÷ 7 = 1

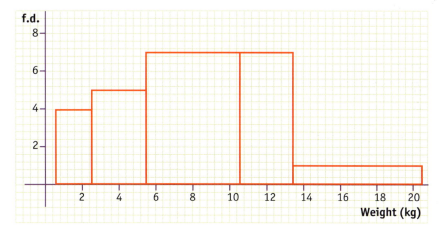

5

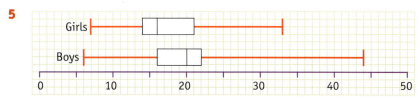

Practice questions

1 (a) $40 \le x < 50$

(b) 4550 ÷ 123 = 37 (rounded)

2 (a) The older you are, the longer it takes you to run a marathon.

(b)

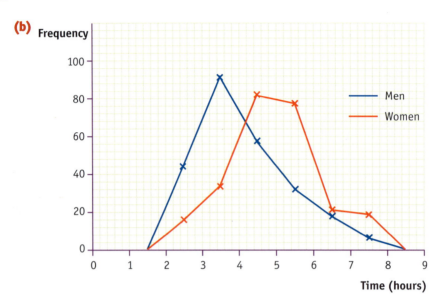

(c) The women's polygon peaks later than the men's, with a flatter distribution. It shows that the women generally took longer to do the run, indicating higher average times for the women.

3 (a)

(b) (i) 95   (ii) 12

(c) (i) James was more consistent because he had a lower interquartile range.

(ii) Helen won more games because her median score was lower.

4

Time	9–10	11–15	16–20	21–30
Width	2	5	5	10
Frequency	2	6	10	32
f.d.	2 ÷ 2 = 1	6 ÷ 5 = 1.2	10 ÷ 5 = 2	32 ÷ 10 = 3.2

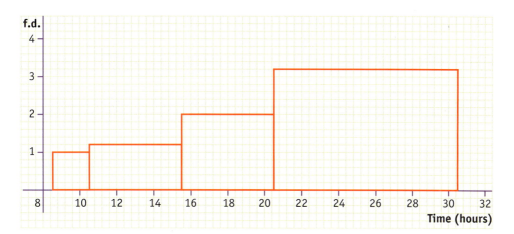

5 **(a)**

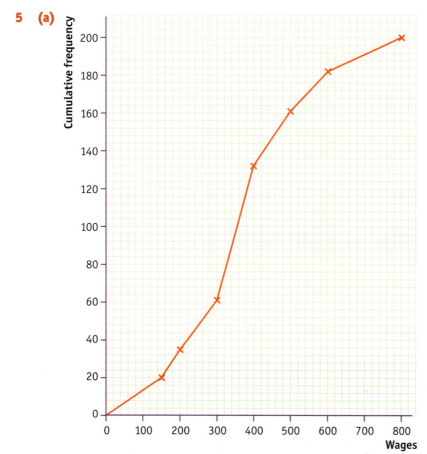

(b) £355

(c) £200

(d) 8

6 (a) (i)

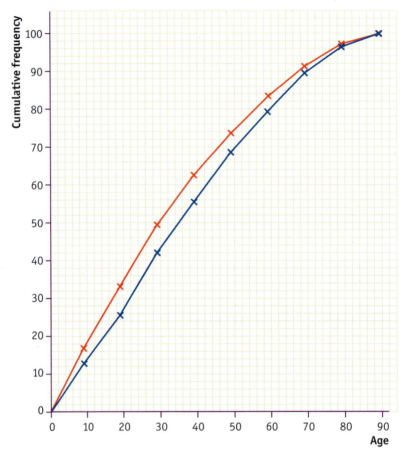

Cumulative frequency (y-axis)

Age (x-axis)

(ii) England 37, Northern Ireland 36

(iii) Northern Ireland has a lower average age than England.

(b) Scotland has a slightly lower average age for males than England, but the distribution of ages is similar.

(c) Northern Ireland has the lowest average age, Wales the highest. England has the smallest proportion of under 20s. The age distributions throughout England, Scotland and Wales are all similar, Northern Ireland having the larger proportion of young people.

7 (a) $0.35^3 = 0.043$ (2 s.f.)

(b) This can happen in three ways: GGB, GBG, BGG
Probability $= 3 \times (0.35 \times 0.35 \times 0.65) = 0.24$ (2 s.f.)

(c) This is $1 - P(\text{no girls}) = 1 - (0.65^3) = 0.73$ (2 s.f.)

8 (a) P(no rain and hold-up) + P(rain and hold-up)

$(0.6 \times 0.25) + (0.4 \times 0.33) = 0.282$

(b) $25 \times 0.282 = 7.05$, hence expect 7 bad hold-ups

9 (a) There are many different ways to do this. One example is:

Do you drive a car?　　　□　YES　　　　　□　NO

(If you ticked NO, then thank you for your help,
you can now hand the questionnaire back.)

What sex are you?　　　□　FEMALE　　　□　MALE

For how many years have you been driving a car?　［　　　　］

How many accidents have you had while driving?　［　　　　］

(b) It is important to get a balance, so you will be looking for things like:
- same number of each sex
- equal numbers of young, middle-age and old people
- some rural dwellers as well as town dwellers
- a good-sized sample — over 200 people